Industrial Relations and the Environment:

Ten Countries Under the Microscope

Volume II

(Reports on Greece, Italy, Netherlands, Spain and U.K.)

EF/93/18/EN

Industrial Relations and the Environment:
Ten Countries Under the Microscope
Volume II

(Reports on Greece, Italy, Netherlands, Spain and U.K.)

ISBN 92-826-6023-0 (Vol. II)
Price (excluding VAT) in Luxembourg: ECU 25

Industrial Relations and the Environment:
Ten Countries Under the Microscope
Volume I

(Reports on Austria, Belgium, Denmark, France and Germany)

ISBN 92-826-6022-2 (Vol. I)
Price (excluding VAT) in Luxembourg: ECU 30

Industrial Relations and the Environment:
Ten Countries Under the Microscope
Volume I and II

ISBN 92-826-6021-4 (Vol. I and II)
Price (excluding VAT) in Luxembourg: ECU 50

 European Foundation
for the Improvement of
Living and Working Conditions

Industrial Relations and the Environment:

Ten Countries Under the Microscope

Volume II
(Reports on Greece, Italy, Netherlands, Spain and U.K.)

Edited by
Andrea Oates and Denis Gregory

Loughlinstown House, Shankill, Co. Dublin, Ireland
Tel: +353 1 282 6888 Fax: +353 1 282 6456
Telex: 30726 EURF EI

Cataloguing data can be found at the end of this publication

Luxembourg: Office for Official Publications of the European Communities, 1993

ISBN 92-826-6023-0

© European Foundation for the Improvement of Living and Working Conditions, 1993.

For rights of translation or reproduction, applications should be made to the Director, European Foundation for the Improvement of Living and Working Conditions, Loughlinstown House, Shankill, Co. Dublin, Ireland.

Printed in Ireland

Contents

Acknowledgements 1

Editorial Overview 3

Country Reports

Greece 7

Italy 41

Netherlands 73

Spain 121

U.K. 177

Volume I contains the reports on Austria, Belgium, Denmark, France and Germany.

ACKNOWLEDGEMENTS

These volumes are the product of many hands. Insofar as the editing process is concerned thanks are due to all the original report authors who generously agred, or at least co-operated, with the editors to amend or clarify their original drafts. Helpful comments were received from Dr Hubert Krieger and Dr Eberhard Schmidt who were both entirely supportive throughout. Particular thanks and praise should be recorded for the efforts of Iris Reynolds at the Labour Research Department and Sheena Anderson, Yvonne Williams and Diana Crayk at Ruskin College for their skill and patience in dealing with the vagaries of transcribing and integrating a bewildering variety of computer diskettes.

Andrea Oates and Denis Gregory
November 1992

EDITORIAL OVERVIEW

Editors of reports which attempt to integrate a number of individual country studies have broadly two choices: They can ruthlessly attempt to fit those reports into a predetermined framework or, they can allow the individual reports more 'freedom to breathe' within a framework which permits the reader to make allowances for national peculiarities but still enables sensible comparison to be made in key areas. We have opted for the latter approach.

The reports which follow were all influenced by the original research design developed by Professor Eberhard Schmidt and Dr Eckhard Hildebrandt. However, such are the differences in legal frameworks, industrial relations structures and processes, and the socio-economic placing of the external environment in the 'realpolitik' of the nation states under study, that individual reports inevitably reflected national idiosyncrasies. Moreover, the reports themselves were initially written by authors who, whilst sharing a common concern for the issue of the external environment and its linkage with industrial relations systems, nevertheless, displayed differing academic backgrounds and experiences. Hence, a common scientific perspective was replaced by a rich mixture of views mainly drawn from the social sciences.

A further complicating factor, insofar as editorial integration is concerned, lies in the differences which evidently persist, first of all, in the formal body of objective knowledge in the subject area. Plainly the linkage between the external environment and industrial relations systems is an under researched area, but that said, some authors were able to draw upon academic studies, whilst others had much sketchier material to work upon. In the second instance, although rising awareness of the problems of the external environment has been a common characteristic throughout the countries studied, the extent to which this has triggered concrete activity at the legislative level, or brought forward a collective response from the social partners has varied considerably.

For all these reasons, it became increasingly clear that attempting to fit individual country reports into a common analytical framework, would run the risk of losing important distinguishing features and experiences.

In the light of the accompanying synthesis reports from Professors Schmidt and Hildebrandt, it is unnecessary to dwell here on a detailed review of the points of similarity and difference which exists in the country reports. However, some brief observations on one or two of the issues which emerge strongly from all or most of the reports are in order.

There is widespread evidence that the 'Social Partners' have been relatively slow to enter the discourse on the potential linkage between the problems of the external environment and the role which industrial relations systems could play in achieving workable means of overcoming these problems. Although employers and union organisations began to see environmental issues as something which needed to be incorporated into industrial policy and practice from the late 1980s onwards, it has been to the various embodiments of the 'green' movement supported by consumer groups and the media that the task of developing public debate has largely fallen.

In part, the relatively slow involvement of the social partners can be explained by the dominance of their traditionally narrow focus on the problems of the working environment. Indeed, many representatives of management and labour could argue, convincingly, that such was the speed and scale of technical change in the 1970s and 1980s and so extreme were competitive pressures that maintaining the internal working environment had to be their overiding priority.

That this narrow focus has proven increasingly unsustainable, has become evident as spectacular pollution incidents, associated with industrial processes, have reinforced mounting public concern at the environmental legacy of years of laissez faire development in Western industrialised nations. The consequences of years of under investment in the external environment by the nations of Central and Eastern Europe has, with recent political reform, also become more widely appreciated.

Faced with these pressures, the social partners have latterly moved towards a search for 'cleaner' technologies or have developed strong marketing policies aimed at repositioning products through an emphasis on their environmental friendliness.

Whilst such moves were widely reported in the country studies, the well established dichotomies of employment protection versus environmental improvement, for employee organisations and profitability versus environmental protection for employers were still exercising considerable influence over the social partners willingness to adopt progressive, but potentially costly (in the dichotomous terms posed) strategies and initiatives.

There was some evidence to suggest that a positive link existed between the nature of the partnership or interaction between employers and trade unions and the propensity to initiate progressive action to limit the harmful impact of production processes. In other words, where a high level of mutual trust and involvement exists regarding planning and co-ordinating change at the workplace, the more likely it is for initiatives aimed at the wider external environment to flourish.

The fact remains, however, that even in industrial relations systems where involvement and participation are highly specified, there appears to be a reluctance to allow employee representative organisations (whether works council or trade union) to be involved fully in the critical areas of decision making. The wider environmental impact of an enterprise is still largely seen as a management responsibility.

Not surprisingly such an approach has tended to confine social partner activity to the level of enabling agreements concerning the wider environment whether at national or sectoral level. Examples, of extensions of such concern to workplace negotiated agreements are seemingly rare.

Two further points with regard to the gap between 'national' and 'local' levels should be mentioned. The first concerns the conflict that can exist between union positions _within_ a particular enterprise, and the position adopted by the _wider_ union movement in the immediate locality. A number of studies mention instances where internally the union appeared to be defending jobs, whilst externally other local unions emphasised the priority needs of the local citizens. At the same time, employers could face a similar dilemma, again identified in the country reports, where employer associations appeared ready to discipline individual members who failed to give sufficient attention to the wider environmental consequences of their particular activities.

The second point concerns the search for and implementation of so-called 'clean technologies'. A number of reports draw attention to the contradictory nature of some 'clean technology' strategies which, on balance, have been shown to be as environmentally damaging as the processes they were replacing. The uncovering of 'half clean' technologies, in a sense, suggests that a technological 'fix' is as elusive as ever. More broadly, it implies that common legislative frameworks and environmental targets which, whilst possibly adding costs to production processes, could nevertheless be applied equally and simultaneously to improve the environmental impact of economic activity must remain high on the political agenda of the European Community. The extent to which the present downturn in the overall economic fortunes of the Community continues to delay addressing this issue, of course, remains to be seen.

Denis Gregory and
Andrea Oates

November 1992

INDUSTRIAL RELATIONS AND THE ENVIRONMENT:

GREECE

by

Dr. Christina Theohari
Mr. Illias Banoutsos

CONTENTS

1. STATUTARY BODIES.................................11

1.1 Authorities for the protection of the environment

1.2 Authorities for Occupational Health and Safety.
....14

2   THE ENVIRONMENTAL AND OCCUPATIONAL HEALTH AND SAFETY
    LEGISLATION...................................... 15

2.1 The industrial relations aspects in the existing
    legislation for the environment

2.2 The industrial relations aspects in the existing
    legislation on OH&S............................ 16

3. PROSPECTS FOR THE ENVIRONMENTAL AND OH&S POLICY
    FORMULATED BY AGREEMENTS BETWEEN THE INDUSTRIAL
    ACTORS... 19

4. STATEMENTS, DEMANDS, PROJECTS AND PUBLIC CAMPAIGNS
    FROM THE INDUSTRIAL ACTORS IN ENVIRONMENTAL ISSUES,
   20

4.1 Employers' organizations and management views

    4.1.1 Employers organization views

    4.1.2 Management views.......................... 24

4.2 Trade union views

5. CURRENT ENVIRONMENTAL CONFLICTS AND THE ATTITUDE OF
    THE INDIVIDUAL ACTORS TOWARDS THEM............... 27

6. SUMMARY AND RECOMMENDATIONS...................... 29

7. ANNEXE... 36

8. REFERENCES....................................... 39

ABBREVIATIONS

L.	Law
M.D.	Ministerial Decision
L.A.	Legislative Act
A.C.	Act of the Cabinet
S.O.	Sanitary Order
O.J.G.	Official Journal of the Government
I.C.	International Convention
L.D.	Legislative Decree
S.C.C.	Supreme Chemical Council
G.S.C.L.	General State Chemical Laboratory
P.D.	Presidential Decree
M.D.	Ministerial Decree
R.D.	Royal Decree
M.O.	Ministerial Order
Circ.	Circular
METPPW	Ministry for the Environment Town Planning & Public Works
MSA	Ministry for Social Affairs
MPPHE	Ministry for Physical Planning, Housing and Environment
Decr.	Decree
Res.	Resolution
M.C.O.	Market Control Order
I.L.C.	International Labour Convention

1 STATUTARY BODIES

1.1 Authorities for the protection of the environment

After the passing of the Law on Town Planning and the Environment (L. 360/1976), many services for environmental issues were formed in Ministries and other Public Organizations. Many public/semi-public and private agencies are currently directly or indirectly dealing with the environment. Specifically, there are nine Ministries, two Legal entities (Regulating Planning Councils for the Environmental Protection of Athens and Thessaloniki) and other public agencies, such as the Greek Tourism Organization, the National Weather Forecast Service, the Greek Marine Research Centre (former IOKAE), the Greek Nuclear Energy Company, the Benaki Phyto-pathologic Institute, the National Observatory of Athens, Thessaloniki and several other cities.

The current responsibilities of the agencies dealing with the environment, will be mentioned briefly.

Ministry for the Environment, Town Planning & Public Works

Its responsibilities, in brief, are the following:
- urban planning, housing and town planning policy making.

- the formation and application of mapping and ownership programs

- the expression of direction and the implementation of governmental policy in environment related issues

- the formation of plans and programs for the application of control and coordination of special programs for environmental protection

- the observation of natural human environmental quality and industrial activity pollution control

- covering of issues not included in the responsibilities of any other agency

- the proposal of legislative measures appropriate to its responsibilities.

After the merger of the former Ministry for Public Works with the former Ministry for Physical Planning, Housing and Environment, the surveillance, supervision, control of studies for land reclamation, water supply and sewage works were included in the responsibilities of the General Secretariat for Public Works.

Ministry for the Industry, Energy and Technology (YBET)

This Ministry has the responsibility for the National Council for Energy and Management of the water and natural resources and "the determination of Industrial policy".

Ministry for Agriculture

Environmental responsibilities include the protection of plants, the legislation and study of issues related to agricultural and non-agricultural activities, the development and protection of water and ground agricultural resources, the management and protection of forests and forest areas, hunting, the development of fishing (estuary, water- cultivation, internal waters) and its protection from pollution and the laboratory testing of waters and soil and legislation on all the above issues.

Ministry for Internal Affairs

Responsibilities involve the watering, protection and control of environmental pollution during the collection, removal and disposal of solid and liquid wastes, at the Local Authorities (L.A.) level.

Ministry for Merchant Navy

Responsibilities comprise the proposal of legislative and administrative measures and the ensuring of the necessary technical background for dealing with marine pollution.

Ministry of Transportation

Responsibilities comprise the proposal, promotion and control of legislative and administrative measures for the environment's protection from exhaust gases, noises and other disturbances from motor vehicle traffic.

Ministry for Public Order

Responsibility is to enforce the application of provisions for environmental protection.

Ministry for Health & Welfare

Environmental responsibility includes the study and monitoring of sanitary provisions and programs for the sanitary protection and cleansing of waters (surface water, underground water, sea water, drinking water) the protection of soil from pollution, the cleansing and promotion of water supply and sewerage systems, the exercise of precautionary and suppressive sanitary control.

Ministry for Labour

Main environmental responsibility is the study, legislation and control of sanitary systems at the work place and workers safety.

Ministry for The National Economy (YPETHO)

This Ministry holds the responsibilities of planning incentives for district and industrial development and controls the financing for development projects and the 5-year Economic Development Plan.

Councils for City Planning and The Environmental Protection of the Cities of Athens and Thessaloniki

The responsibilities of these councils include the observance and application of regulating plans of the major areas of Athens and central city plans and harmonizing with those of the environmental protection programs and other agencies. The application of protection measures and control of works and activities affecting the environment. In fact, environmental policy is determined mainly by the Ministry for the National Economy and the Ministry for Agriculture. The role of the Ministry for the Environment, Town Planning and Public Works is to make proposals for environmental issues. Essentially it has no power to exercise precautionary or suppressive control to activities or acts affecting the environment. The main characteristics of the administrative mechanism or legislative framework for the environment is the sub division of the responsibilities among the related Ministries. This, it is said leads to a lack of collaboration and coordination among the respective authorities, an absence of specialized executive staff and an inadequate exercise of inspection and surveillance by the public services. In consequence, it has been argued that the environmental dimension does not have a high profile or sufficient influence on public policy.

The Technical Chamber of Greece

The Technical Chamber of Greece (TEE) is by Law the state's official consultant. It is thus directly involved in development issues of all kinds, such as industrial development, technical infrastructure development, environmental planning and protection, etc.

1.2 Authorities and Institutions for Occupational Health and Safety (OH&S)

The main authorities for Occupational Health and Safety (OH&S) in Greece are:

- The Ministry of Labour

- The Ministry for Industry Research and Technology (YBET)

- The Foundation for Social Insurance (IKA)

The Ministry of Labour is the authority for inspecting the working environment in all sectors of economic activity, except for mines and quarries. It comprises the Directorate for Working Conditions, responsible for policy making and coordination of the Labour Inspectorates on prefectural level and the Centre for Health and Safety at Work (KYAE) which provides the scientific and technical support to the inspectorates.

The Ministry for Industry (YET) is responsible for the inspection of mines and quarries.

The Centre for Occupational Medicine and Work Physiology which belongs to IKA is responsible for the investigation of occupational diseases. Apart from the state authorities which play a regulatory role, there are other organizations which have a strong influence on OH&S issues. These are the Technical Chamber of Greece (TEE) with an advisory role as mentioned earlier, professional organizations (occupational physicians, ergonomists, etc.) and trade union organizations mainly the Greek General Confederation of Labour (GSEE) and the Athens Labour Centre (EKA), the last being the only non-public organization with a bureau on OH&S employing a full time professional ergonomist and a part time occupational physician. The framework Law 1568/85 for the health and safety of workers has introduced three important institutions at enterprise level for the promotion of the improvement of working conditions. They are exclusively advisory and are:

- The Health and Safety Committee, which is described in more detail in section 2.2.

- The safety engineer, and

- The occupational physician.

According to a transitional provision included in L.1568/85 these three institutions apply only to enterprises employing over 150 persons. Other institutions introduced by the L.1568/85 are the Council for Health and Safety at Work (SYAE) at a national level and the Prefectural Councils for Health and Safety at Work on a regional level, again both are advisory.

2. THE ENVIRONMENTAL AND OH&S LEGISLATION

2.1 Industrial Relations aspects in the existing legislation for the environment

Central works councils (KES) operate within public enterprises, namely the Public Power Corporation (DEH), the Hellenic Railway Corporation (OSE) and the Hellenic Telecommunications Corporation (OTE). These councils were formed at the enterprise headquarters and established after a decision of the Minister for Energy and Natural Resources. A central works council has the following responsibilities:

- to introduce issues regarding the organization, operation, planning and control of the enterprise,

- to propose methods that increase the productivity and improve the enterprise services,

- to give opinions to the Supreme Council for Social Control (S.C.S.C.) on regulatory issues and to the Board of Directors about the application of regulations,

- to give opinions on matters concerning working conditions health and environment,

- to decide about the formation of local work councils whose composition and responsibilities are similar to those of KES.

- to decide its operating plan and methods.

- to decide about cultural, athletic, personnel clubs and restaurant issues, financed by the enterprise budget.

A typical KES has nine members and is constituted of employees' representatives directly elected following a general directly proportional voting system. The members tenure of office is three years.

These central works councils can contribute to the active intervention of employees on environmental issues resulting from the operation of these three public enterprises.

Note: Before the constitution of KES experimental works councils were formed in Public Power Corporation (DEH) according to a collective agreement in 1983.

A Presidential Decree of April 6th 1988 - (O.J.G. 63), implemented the 135th International Labour Convention referring to the formation of employees councils of enterprises.

This Agreement determines those responsibilities of the employees council that do not refer to issues related to the external environment. These responsibilities are limited to the health and safety regulation of the enterprise, i.e. they focus on the working environment.

2.2 Industrial Relations aspects of the existing legislation on OH&S

There are two basic laws in Greece having a direct impact on industrial relations with respect to OH&S. Firstly, the Law 1568/86 "Health and Safety of the Workers" and secondly the Law 1767/88 "Work Councils and other Labour Regulations - Ratification of the 135th International Labour Convention".

The two main participatory bodies for OH&S on enterprise level are the health and safety committees and workers Councils provided by the Laws 1568/85 and 1767/88 respectively. According to the Law 1568/85 (article 2a, paragraph 1) "The workers employed in enterprises with more than 50 employees have the right to set up Health and Safety at Work Committees, consisting of their elected representatives in the enterprise. In enterprises employing from 20 to 50 persons, a Representative for health and safety at work is elected". However, according to the transitional period defined by the same law (article 14, paragraph 1), a health and safety Committee may be elected in enterprises which employ over 150 persons.

Law 1767/88 gives the right for workers to elect work councils in undertakings with more than 50 employees. This right extends to undertakings with more than 20 employees if there is no other kind of trade union in the enterprise. The work councils have broader responsibilities than the health and safety committees and enjoy the right of common decisions with the employer on certain issues. They also appoint the members of the health and safety committee.

The health and safety committee, either elected directly or defined by the work council, has the right to be informed about:

- Industrial accidents and occupational diseases occurring in the undertaking.

- The introduction of new technologies in the enterprise, new production processes and substances, insofar as these measures affect health and safety conditions.

- Any harmful agents the workers are exposed to.

The work council has the right to be informed, before any final decisions are taken, on the following matters:

- Yearly expenditure plans on measures for health and safety at work.

- The undertaking's policy for health and safety.

- Any change of the plant's installations, like relocation, expansion or shrinkage.

The health and safety committee's advisory rights are:

- To propose measures for the improvement of working conditions.

- To propose measures to prevent occupational accidents and other work related hazards.

- To call upon the employer to take all appropriate measures in the event of imminent or serious hazard, without excluding the stoppage of machinery, installation or production process.

The committee or its representative may call upon an expert for health and safety at work, with the consent of the employer.

The work council also has the right to propose measures for the improvement of working conditions.

Health and safety committees hold meetings with the employer within the first ten days of each three month period in order to resolve issues which have arisen in the enterprise related to health and safety at work. Minutes of meetings are written in duplicate and one copy is given to the committee. The work council may meet with the employer more frequently, once every two months. These joint meetings of both the health and safety committees and the work councils with the employer are considered as being an important form of negotiation.

The work council also has the right to common decisions with the employer on the undertaking's policy of health and safety at work.

In building sites and in ship-building and ship-repairing activities where, due to the small number of employees neither committees nor councils may be elected and also because of their inherent high occupational risk, joint inspection committees for health and safety have been instituted by Ministerial decision. These joint committees consist of two union representatives, the labour inspector and a representative of the Technical Chamber of Greece. In the joint inspection committees for ship-repairing and building activities there is an additional member representing the port police authorities.

On a national level, Law 1568/85 institutes the council for health and safety at work (SYAE). This is an exclusively advisory body consisting of representatives of the employers, employees, professional organizations and experts. On a regional level similar advisory bodies are established (NEYAE), to coordinate the relative activities and to initiate others.

Members of the work councils have the right to twelve days paid time off for training which may include topics on health and safety as well as environmental issues.

3. PROSPECTS FOR ENVIRONMENTAL AND OH&S POLICY FORMULATED BY AGREEMENTS BETWEEN THE INDUSTRIAL ACTORS

To date, there have been no agreements between the industrial actors concerning the environment. The only reference made is on OH&S issues.

On January 1990 an agreement was reached between the Greek General Confederation of Labour (GSEE) and the Federation of Greek Industries (SEV). This agreement covered topics like rights of health and safety committees, duties of the enterprises with regard to the appropriate health and safety resources, common health and safety at work demands (GSEE - SEV) to be fulfilled by the State, and finally a basic outline of the National Institute for Health and Safety whose establishment was considered to be a high priority. It must be noted that one of the main objectives of this future Institute would be the investigation of the inter-relationship between the inner and outer environment. This agreement, although it had been signed by both the groups of experts nominated by GSEE and those nominated by SEV, has never been ratified by the two sides nor has any attempt been made to promote it in practice.

The General Collective Labour Agreement 1991-1992, which was signed on March 1991, mentioned in article 6 that the GSEE on one side and the three employers' organizations on the other side (SEV: Federation of Greek Industries, GSEBE: General Confederation of Self-Employed and SME's owners, EESE: Federation of Unions in Commerce) would establish a bipartite Institute for Health and Safety at Work. Again, one of the main activities would be to explore the working environment matters inter-related with the external environment.

A decision whether this Institute would go ahead or not, was due by the end of 1991 the main responsibility lying with the Minister of Labour who by law ratifies this collective agreement.

The premium which the employers organizations have agreed to pay yearly per employee is considered to be more than adequate to finance the Health and Safety Institute. Its activities, apart from research will include training of employers and employees as well as information and publications on OH&S and environmental issues filling a real and crucial gap and creating the conditions to improve the dialogue between the social actors on these issues.

4. STATEMENTS, DEMANDS, PROJECTS AND PUBLIC CAMPAIGNS FROM THE INDUSTRIAL ACTORS ON ENVIRONMENTAL ISSUES

4.1 Employers Organisations and Management views

In the absence of any policy statements and demands by the employers side on environmental issues, the only way to present their views was to interview representatives of employers organizations and to hold discussions with the management of certain enterprises.

4.1.1 Employers Organisations views

The president of the Association of Industries of Attiki, Mr. Christos Fyrogenis, was interviewed using a structured schedule of questions which covered the following:

- What actions have been made in order to inform the employees or workers representatives on issues regarding environmental protection from the production processes?

- Have the employers' organizations ever expressed in public the management's views on environmental protection by means of declarations or expressing their opinion on serious legislation matters or by any other means?

- Do the enterprises employ environmental specialists? Do the enterprises have a policy of developing close relationships with consumers organizations or residents complaining about pollution issues?

- Have the employers' organizations carried out studies or opinion polls to investigate the awareness of the management with regard to environmental protection?

Mr Fyrogenis agreed that the environment and the problems caused by human activity, have become part of everyone's vocabulary in recent years. For industry the problem has always been there, although today it appears more intensively, because industrial activity has to face the environment as it really is, i.e. a precious, essential and scarce resource.

Drawing on the experience of Attiki and its industries, Mr Fyrogenis pointed out that the above mentioned position had been adopted many years ago and had been declared officially so by the Association of Attiki Industries, earlier than any other district based industrial association. Moving from the stage of denying the existence of the problem to the point where it was accepted and ways to smooth the adverse environmental consequences of industrial activity were being explored was a position not easily reached and was not without some reactions. It was nonetheless essential for the more modern and advanced industries to give a good example, to emphasize related issues and to call for a steady and continuous intervention by the Association of Industries of Attiki in order to challenge the view that environmental protection can be achieved only in theory. The Association's efforts were needed to convince its members that environmental protection carries an infinitely smaller cost than environmental devastation, and that it is an essential parameter of economic development and cannot be ignored.

To emphasise these facts, the Association of Attiki Industries, has intervened and still does, trying to persuade industries that they can and should go beyond adopting minimum environmental protection measures. Within this framework and to indicate clearly their position the Association of Industries of Attiki has :

- Denied support to members of the Association that were reluctant to apply measures imposed by the state.

- Promoted the reinstallation of sections of Industrial activity, when this was essential for environmental protection reasons.

- Adopted the use of mild forms of energy whilst continuously pressing the state towards accelerating the rate of use of natural gas.

- Presented strongly the view that industrial modernization is closely related with the investments necessary for the protection and upgrading of the Attiki basin environment.

- Convinced the industries of Attiki, by means of circulars, leaflets and interventions, that environmental protection is not just an obligation, but also the strongest arguments against those who believe that industrial development is by nature contradictory to environmental protection. This resulted in industries continuously lowering their participation in the environmental degradation of the Basin, in contrast with other activities in the area (automobiles, central heating, public activities, etc.) who continuously increased their degradation effect.

According to Mr Fyrogenis, the following facts were characteristic and proof not only of the intervention of the Association of the industries of Attiki, but were of interest in their own right insofar as environmental protection was concerned.

- There are industries which, although not obliged, recycled their liquid wastes and used them to water gardens and alleys they had developed in their surrounding area.

- Similarly there were industries who every summer, although again not obliged to, nonetheless decreased their activity or prolonged voluntarily, periods of low activity, when the conditions demanded it.

Mr Fyrogenis agreed that environmental problems caused and still cause reactions, positive or negative, which vary according to the degree of their objectivity, and their realistic or non realistic attitude towards the problem. He pointed in particular to:

- The press and the fact that it had become more impartial in dealing with various polluting sources, and in the presentation of the positions, opinions and commitments of industries in this regard.

- The extent to which public opinion was misinformed and totally negative to industry until a few years ago, but had finally begun to face the truth and understand that environmental protection demands strong and modernized industry.

- Local Authorities (although luckily not all of them), who by contrast appeared to be the main opponents to the existence of industry and to its modernization and environmental protection efforts. Although the local authorities should have adopted a policy of attracting industries instead of opposing them, many evidently did not accept that industry and residents can co-exist. The fact that the existence of viable industrial units, meant a growth in income and a lowering of unemployment rates for the municipalities did not seem to matter. In order to reverse this negative attitude, the Association of Industries of Attiki tried to inform municipalities about the measures it adopts and its commitment to environment protection. The formation of common-trust committees for the inspection of the application of environmental measures was accepted, and the Association participated in the committee of the Municipality of Athens formed to deal with air pollution.

In contrast, mention should be made of the relative reluctance the state had shown (until recently) in adopting the proposals for the modernization of the industries of Attiki. The Association intervened with studies, data and memoranda, to prove that the environment of Attiki could only be protected with modern industry, and that industrial investments exclusively made for environmental protection, would total many billions of drachmas.

Recently, and Mr Fyrogenis felt this to be clearly positive, the opinion of the people working in industry, had begun to be heard. He felt that people working in industry today realized that the economic development of the country and their employment depended on industries still remaining in the Attiki basin which in turn depended on the adoption of measures for environmental protection as proposed by the Association.

Finally, Mr Fyrogenis said that today more than ever before, the conditions for the modernization of the industries of Attiki, in parallel with the adoption of environmental protection measures, were optimistic. This was not only because of the EEC and its legislation but rather that industries knew they could not operate without environmental protection and were therefore determined to undertake the investment to achieve this purpose.

4.1.2 MANAGEMENT VIEWS

With regard to a typical management view of environmental protection issues, the Aspropyrgos Refineries (ELDA) were asked to give an extended and detailed briefing on the way the environmental dimension is treated in its policy.

This enterprise had made significant efforts in the direction of reducing the environmental effects of its operation. During recent years significant modernization investment had been made, a large percentage of which referred to the adoption of anti-pollution technology. It should be noted that 25% of the total investment for new installations was spent on environmental protection measures.

A special research and development department operated within the enterprise, dealing mainly with environmental quality control and energy conservation. The monitoring of environmental quality, was achieved by means of an air pollution measurement station, a waste water treatment plant and proper toxic waste disposal.

Energy conservation concerned techno-economic issues regarding the optimization of the operation of the plant so that energy losses were minimized. The appropriate operational research was carried out in this direction.

The Environmental Department of the enterprise in cooperation with the Public Relations Department took care of information bulletins which mentioned the actions of the enterprise for environmental protection.

Apart from ELDA other enterprises exhibited environmental sensitivity expressed in various ways. These ranged from "ecologic" advertisements, product advertisements made through or in parallel with environmental issues, to the publishing of bulletins or even books about environmental enterprise activities.

These examples apart, there was still a long way for enterprises to go in order to realize and contribute to environmental protection more effectively.

The initiatives of some Greek enterprises whilst encouraging could not be considered adequate.

4.2. TRADE UNION VIEWS

In order to obtain a clear view on the issue, the president of the Greek General Confederation of Labour (GSEE) and the president of the Athens Labour Centre (EKA) were interviewed (using a structured question schedule) since they represent the workers, the first at a Panhellenic level and the second for the Athens major area.

Mr Lamros Kanellopoulos the President of the General Confederation of Greek Labour said that, "The environmental problems in Greece were very serious and had continuously worsened."

The General Confederation of Greek Labour (GSEE) believed that this deteriation was very important both from the point of view of the workers' quality of life as well as for the preservation of the environment itself.

He stressed that both large industries as well as small and medium size enterprises appeared unwilling to invest in the technological modernisation which was related to a large extent to antipollution measures.

In Mr Kanellopoulos's view Governments in Greece were responsible to a large degree for failing to pursue the appropriate measures for the necessary modernisation of the country's production infrastructure in accordance with environmental protection principles.

Even the legislative framework for the establishment, extension and operation of industrial activities (cost-benefit studies, studies on environmental impact assessment, etc.) had been neglected for fear of inhibiting industrial investment.

The tendency to look for easy and immediate profits had been accepted for too long at the expense not only of the environment but also of the future of the industrial infrastructure of the country.

Mr Pan Ploumis the President of the Athens Labour Centre pointed out that the Athens Labour Centre made efforts to deal with environmental issues on a permanent basis. For this reason, a special department had been formed, staffed with an environmental engineer. A member of the Executive Committee was also assigned to observe these efforts.

Mr Ploumis felt that a lot of effort still had to be made to convince trade unions that environmental issues were a section of work that directly concerned them and that they should actively intervene.

He agreed that this would not change trade union characteristics, or divert them from the purpose they had been established to achieve. On the contrary, their targets would widen and develop, as would their contact with parts of society sensitive to these problems. In so doing unions would be seen as organizations capable of facing modern problems, dealing with them and protecting workers from the consequences of environmental crises. The Athens Labour Centre focused its attention and developed activities in the following areas:

- The study and application of positions and proposals made on different issues such as: industry and its contribution to environmental pollution, car usage and air pollution in Athens. These studies were made with the support of specialized scientists, other agencies (mainly scientific), and presented and discussed in meetings, seminars, conferences, etc.

- Organising meetings and submitting proposals to the Ministers responsible for environmental protection.

Such meetings were used to pressure for certain environment upgrading measures to be taken and applied.

The Centre often issued reports to newspapers, TV and radio on specific problems caused in the Athens area.

The Athens Labour Centre also made great efforts to advance knowledge and sensitivity on environmental issues. Many seminars had been organized and more than one hundred trade union members had participated.

The Centre at the same time had developed relationships with individuals and social groups that act in this field. Close co-operation with the Technical Chamber of Greece (TEE) scientific associations, such as the Panhellenic Association of Chemical Engineers, Mechanical-Electrical Engineers, the local authorities and peripheral municipalities, had been established and links formed with ecologic teams interested in the protection and development of grassland and, the upgrading of urban spaces and country areas.

Mr Ploumis held it to be a fact that contradictions, in fighting air pollution and protecting the environment on one side and maintaining a factory so that workers maintain their jobs on the other, had only been evident on a few occasions. The Athens Labour Centre aimed to solve these problems on the basis of the disposition of the necessary resources for the modernization and improvement of of enterprises, at the same time taking certain measures for the protection of the environment (use of filters etc.) The Centre agreed that the necessary finances should be given to employers so that they could transfer their enterprises to specially developed areas away from residential ones, as in the case of tanneries and other major polluting industries.

5. CURRENT ENVIRONMENTAL CONFLICTS AND THE ATTITUDE OF THE INDUSTRIAL ACTORS TOWARDS THEM

Quite often, the sharpening of environmental problems has lead to strong protests and disputes between small and medium enterprises (SME's) and residents of nearby areas. This happened when:

- The SME's operated under conditions that did not comply with the regulations for environmental protection. These conditions lead to degradation, pollution and even destruction of the environment with direct impact on the residents' health and life. As a large proportion of production activities coexist with residential areas, (resulting from the lack of proper planning in the past), friction was frequently caused. An important contributing factor was the lack of trust towards state control and efficient enforcement of the laws in combination with the almost nonexistent role of Local Authorities in inspection procedures.

These facts limit alternative solutions for some industrial units to be modernized in situ especially because this in most cases means extensions, although for some units it may mean a relocation to an other area.

Whereas economic crises tend to put residents of industrial regions in a dilemma: the fear of possible job loss conflicts with the process of enforcement of environmental protection rules. The worsening of environmental conditions, in combination with inadequate social infrastructure, has, from time to time, finally lead them to overcoming the above mentioned fears and to fight for their right to live in a clean environment.

There are areas in Attiki, that have literally been suffocating due to the concentration of industries in a limited space and to the absence of infrastructure necessary for environmental protection.

Reference in this respect is made to Drapetsona, Keratsini, Aspropyrgos, and Elefsina. In these areas residents were particularly sensitive and active for the enforcement of the law.

These reactions of the residents have been expressed in a more organized way with the help from the local authorities. A recent example of this was the opposition of the residents of Elefsina to the extension of the existing refinery plant. When the "Panelefsinian Front" (municipalities and agencies of the town) conducted a ballot, the extension was condemned.

Such campaigns have taken place in other areas too, in Attiki and throughout the country campaigns have been organised to prevent actions that are considered harmful for the environment.

It must be noted though, that there have been cases when exaggeration, suspicion and prejudice to any investment occurred. This situation can be avoided through the active participation of the enterprises' management, informing people of the measures they take for environmental protection.

6. SUMMARY AND RECOMMENDATIONS

Industrial development and the environment

The answer to the question, whether environmental issues have been introduced into the system of industrial relations in Greece, is negative.

A number of causal factors have contributed to the non existence of such issues within the present industrial relations system. These causes are of decisive importance and are related to the political, economic and developmental choices adopted in Greece during the last decade. These choices have not contributed at all to the introduction of demands for the protection of the environment within the bargaining framework of the workers.

The lack of policy for industrial development in accordance with the objectives of protection has a negative effect on the attitude of employers and workers towards the introduction of environmental issues in industrial relations.

In general, it can be said that the relationship between industrial development and environmental protection in the country, is directly related to:

- technology level

- the spatial distribution of industrial activities

- the environment protection attitude of the production partners.

a) Technology level

It is obvious that technological progress can greatly contribute to a drastic decrease of pollution caused by production processes. The adoption of "clean" production processes during which the products are made at the lowest possible environmental cost is one pathway to the lowest possible pollution. Energy and material saving, will also contribute to the reduction of negative effects and the elimination of prejudice as well.

It should be noted that in "modern" technologically advanced factories pollution protection devices are incorporated in the initial design of the production process.

Taking this as given, industrial modernization, which is a kind of industrial development in developing countries, can co-exist with environmental protection. Greece is in that development stage during which the modernization of production processes is of utmost importance. If this is accomplished in line with environmental protection principles, then the question about the compatibility of industrial development and environmental protection will be answered positively both in theory and in practice.

b) The spatial distribution of production activities

The spatial distribution of production activities throughout the country involving the location of residential areas on one side and the appropriate infrastructure facilities for their operation (roads, port, transportation nodes, etc.) on the other, plus the observance of environmental rules during the choice of location of industrial units as well as their effective inspection (which is the state's obligation), are basic factors for proper industrial development. Such a balance would also lead to the minimisation of prejudice which has, reasonably in many cases, characterised the general public's attitude.

Peculiarities have led (and still lead) to environmentally degraded conditions that encourage an a-priori negative attitude to any investment initiative that may appear in Greece. This attitude may be attributed to the state's incapacity or unwillingness to control effectively production activities for fear of it constraining the arrival of new investments. This tactic, though, has led to a delay in the modernization of production means with consequences on industry, small and medium sized enterprises, production in general and to uncontrollable environmental degradation, especially in areas with high densities of production activities. Such a case is that of the Attiki basin where 47% of the country's industry has been concentrated without the observance of town planning and environmental criteria.

In many cases, the in situ modernization of many of the industrial units that are spread throughout the residential area is impossible. The fact that production activities are dispersed throughout the residential web by itself creates pollution. Even those units that do not pollute, cause serious disturbances such as traffic burdens. The development of Industrial Areas and Industrial Parks has been realized partly as a policy of decongestion leading to the creation of conditions for the proper functioning of industrial activities and environmental protection. Industrial Areas today exist in 19 prefectures while another 10 have been planned. The possibility of Industrial Areas being established in all the country's prefectures has been raised. The basic concept of Industrial Areas, whose implementation has been undertaken by the Greek Bank for Industrial Development (ETVA), is that they constitute the National Framework of Industrial Development Centres, a network developing near urban centres of the country.

c) Environmental culture

The development of an environmental conscience in all production actors is an essential factor towards the enhancement of environmental protection conditions.

This conscience is acquired through education at school, mass media, scientific agencies, syndicates and political centres of the country.

ANNEX

THE LEGISLATIVE FRAMEWORK CONCERNING THE ENVIRONMENT IN GREECE

With reference to Greek legislation on the environment the following paragraphs, are crucial in providing an integrated picture of the institutional framework.

Greek legislation for the environment began in 1912 currently it includes more than 800 Legislative Acts, i.e. Laws, Presidential Decrees, Royal Decrees, Sanitary Orders, Ministerial Orders,. Acts of the Cabinet, etc., related, directly or indirectly, to the protection and management of sectors of the environment and to activities, or policies that affect it.

The legislative and administrative framework for the environment is made up of laws that are divided in the following groups :

a) Constitutional Orders

b) General Institutional Laws for the environment

c) Agencies' Responsibilities

d) Physical Environment - Nature protection

e) Town Planning

f) Pollution

g) Energy

h) Chemical substances

i) Industry

j) Associative procedures - Local authorities

k) International Conventions

l) EEC Legislation

m) Approved 5 year Programme for Economic Development

CONSTITUTIONAL PROVISIONS

Article 24 of the Greek Constitution of 1975 provides that the state is obliged to protect the physical and political environment and must impose special preventive and repressive measures in order to fulfill these obligations.

Town planning and protection of traditional regions and elements fall within the meaning of the term physical environment whereas, the political environment includes all human creations, i.e. monuments, works of art, traditional regions and ancient ruins.

All necessary measures taken in order to prevent the disturbance of the environment and protect it from any hazards are included in the term protection of the environment. As a result, all human actions that could make changes to nature, thereby jeopardizing the survival of human kind, are also included in the protection of the environment.

GENERAL INSTITUTIONAL LAWS FOR THE ENVIRONMENT

There are three important laws which create the Institutional framework for the protection of the environment. The "Town Planning and Environment" law passed in 1976, set up the National Council for Town Planning and Environment (NCTPE), which comprises the Prime Minister and ten Ministers (Coordination, Economics, Agriculture, Civilization and Sciences, Town Planning - Habitation and Environment, Industry and Energy, Internal Affairs, Social Services, and Commercial Marine). Other Ministers and Deputy Ministers in charge may be called to participate in the NCTPE as well as representatives of local authorities, public organizations and enterprises.

The NCTPE's responsibility is to make decisions about town planning topics, supervise the application of appropriate programs and coordinate the work of the implementing agencies.

In 1980 an important law was passed establishing a Ministry for Physical Planning, Housing and Environment with the authority to make special town planning studies and programs (except for national town planning), control the application by supplemental agencies of a special programme for environmental protection, determine habitation policy and prepare and apply programmes and land registration.

Additional responsibilities were given in 1982 to the Ministry concerning the co-ordination of the Environment Directorate from the Ministry of Coordination, and increasing its role in town planning and environmental protection.

More recently, in 1986, a law for environmental protection was passed.

The scope of this law is to establish fundamental rules, criteria and mechanisms for environmental protection, so that man, as an individual and as a member of society, lives in a high quality environment, in which health is protected and the development of personality is favoured.

LEGISLATION FOR THE ENVIRONMENT

Specific legislation exists to provide control in the following areas:

Air pollution

Drinking water

Marine pollution

Soil pollution

Fuels

Chemical substances - preparations

AIR POLLUTION

Since 1912, more than 140 peices of legislation including laws, Presidential Decrees, Ministerial Decrees and Acts of the cabinet have been passed aimed at controlling air pollution in Greece. Whilst many of these legal instruments have concerned themselves with the mounting problem of pollution associated with car usage, a number have plainly been enacted to comply with European Community Directives.

LEGISLATION FOR DRINKING WATER

Since 1943, at least 12 individual legislative instruments have been passed aimed at controlling the quality of drinking water and protecting it from pollution. Recently, attention has focussed on the protection of water resources from contamination by toxic substances and liquid waste. Again, the need to harmonise with EC directives and guidelines in this respect has driven recent changes in Greek legislation.

MARINE POLLUTION

As may be expected in a country with such an investment in Merchant shipping and tourism, the need to control and reduce marine pollution has been a subject for legislative action since 1991. Some 27 separate actions can be identified aimed at fulfilling these objectives.

SOIL POLLUTION

This appears to have received relatively little attention save for a ministerial decree of 1986 on Solid Waste which provided for harmonisation with an EEC directive.

FUELS.

Legislation here, mostly by Ministerial Decree has attempted to control Petro chemical and gas installations whilst also promoting the use of lead free petrol by motorists.

CHEMICAL SUBSTANCES - PREPARATIONS

Some 15 Ministerial or Presidential Decrees have been adopted mainly in the 1980's aimed at restricting the use of certain dangerous substances e.g. pesticides. Again EEC directives have proved to be important in driving change in these areas.

NOISE

Since 1950 Greek law has recognised the need for citizens to be protected from excessive noise levels. Recent Ministerial and Presidential Decrees have been focussed on noise generated by aircraft and motor cycles.

LEGISLATION CONCERNING INDUSTRIAL ACTIVITY

The basic legislation which provides for the granting of permission to establish industries that were dangerous, obtrusive or unhealthy was passed in 1912. The system of penalties for those companies who broke the law was considered adequate at the time. A departmental Advisory Council was set up in 1962 to advise on the application of this original legislation. Subsequent legislation in the 1960's further widened the scope for technical control over potentially hazardous installations and processes. Towards the end of the 1960's, in an effort to speed up the planning processes attention was placed on faster systems for granting permission.

By the 1980's the first attempts at including assessments of environmental impacts within the planning framework began to appear. A presidential decree of 1981 scratched the surface of the need for environmental protection by requiring environmental impact studies to be submitted as part of any plans for new investment or extension of existing capacity. The inadequacies of this particular decree have been identified as follows:

- the study for the disposal of liquid wastes approved by the prefectural sanitary authorities which are generally considered to be totally insufficient, is incorporated in the Environmental Impact Study. (S.E.I).

- the specifications for the S.E.I. cover the stages of approval and establishment only and not those of construction and operation

- no reference is made to environmental quality criteria

- no provision is made for the decentralization of responsibilities and the local authorities and other relevant agencies participation in the S.E.I. and pollution control approval procedures

- it provided for opinion taking from the Ministry for Industry but did not determine the responsibility of other agencies that, according to the legislation in power, give their opinion on the impact of such activities on natural resources, water and the cultural and historical environment.

More recently, in 1986 legislation "For the environmental protection" included provisions for the categorization, approval, observance and control of of the relevant terms of environmental impact assessment projects, and the establishment of environmental quality and emission standards, together with the operation and maintenance of waste treatment plants.

It must be pointed out though that, for the realization of these provisions, the issue of a number of Ministerial Decisions or Presidential Orders is necessary. To date there has been no evidence of such action.

ECONOMIC INCENTIVES

As can be seen from the relevant legislative provisions, in the interests of incentives, the procedures that could have helped protect the environment from the effects of industrial activity, have been simplified or abolished. This abolition has been accompanied by the financing of private initiative from the state budget and the inability to exercise essential regulatory policies.

For example, a law passed in 1981 "for the offering of incentives for the district and economic development of the country and the settlement of relevant matters", provides the ability to simplify the processes for the issue of installation and operation permission for the establishment, extension, rearrangement of industries, handicrafts and any kind of mechanical installations and storehouses.

Similarly, a law passed in 1983 "For the improvement of investments, organization of State Procurements Services and other provisions" in effect accelerates the procedures for issuing permissions for installation or extension or modernization of industries and gives the Minister for National Economy the right to issue permission, after taking opinion from the responsible Ministry within 40 days.

PARTICIPATORY PROVISIONS

Legislative provisions, provide an involvement of local authority (L.A.) and employees' representatives, in environmental issues, regarding decision making and control exercising.

- A Presidential Decree of 1982 determines, among other things, the responsibilities and obligations of municipal and communal special service personnel regarding the following

- cleanliness, traffic and parking of vehicles, building, environmental pollution, protection of ground-water potential.

Given the local authorities' (L.A.) harsh financial conditions, it has proved impossible to meet the above provisions and local authorities are just limited to cleanliness work.

Advisory responsibilities of municipal Councils, are set out in legislation passed in 1982 in the following areas:

- water supply and sewage in all common use networks

- protection of the natural, built and cultural environment

- traffic and transportation of the district or region

- urban planning development and recreation of the region

- construction of new works and maintenance of existing works

Prefectural Council

With regard to environmental impact assessment projects the Council must inform the citizens and their representative agents before approval is given. The way of briefing is to be determined with a Ministerial Decision, which has not been done yet.

The Council gives its opinion about Government plans regarding the categorization of regions, elements or entities of the natural environment and landscape and the determination of borders and possibly protection zones.

It participates, with five members, for each case, in the Regulating Plans Councils for the Environmental Protection of Athens and Thessaloniki.

Local Authorities (L.A.)

Among the environmental control responsibilities which LA's possess are the following:

- They check the documentation and approve the environmental terms, after the decision of the Mayor or Officer in charge, for works or activities having particularly obtrusive or dangerous consequences for the environment.

- They undertake the execution of works, that are in accordance with the approved environmental terms for the execution of works and activities.

- They give their opinion to the Prefecture in charge for the determination of areas, where the final disposal of domestic waste is allowed.

- They participate in Regulating Plans Councils for the Environment of Athens and Thessaloniki with representatives of the relative Municipal Council and Municipalities and Communities Union.

Other Agencies

The participation of one representative of agencies to the Regulating Plans Councils for the Environmental Protection of Athens and Thessaloniki, is provided for. These agencies are the following:

> The Geotechnical Chamber of Greece, the Commercial and Industrial Chamber of Athens and Thessaloniki (the city interested is participating each time), the Economic Chamber, the Art Chamber, the Higher Civil Servants Directorate Committee, the Greek General Confederation of Arts and Crafts, the employees of Social agencies organizations and syndicates.

References

1. Atmospheric pollution in the Athens area Technical Report Ministry for the Environment, Town Planning and Public Work EART Department - PERPA Athens, January 1989

2. New Hydrology Journal Issue no. 84, October 1991

3. Ikonomikos Tachydromos Journal Issue no. 36, September 5, 1991

4. Environment and Development Technical Chamber of Greece (TEE), Athens 1988

5. Industrial and handicraft activities and environment EKA Publications, Athens 1991

6. 2nd Conference on Industry - Greek industry perspectives TEE Publications, May 1989

7. VI.PE, Industrial Zones of ETVA ETVA Publications

INDUSTRIAL RELATIONS AND THE ENVIRONMENT

ITALY

by

Dr. Alessandro Notargiovanni

TABLE OF CONTENTS

        Page

1. INTRODUCTION - The Development of Environmental Issues and Industrial Relations.................. 44
2. THE LEGAL FRAMEWORK............................. 47
2.1 Introduction
2.2 Environmental legislation
    2.2.1 Areas with a high risk of environmental crisis................................... 48
    2.2.2 Draft agreement between the Environment Ministry and the trade union federations.. 49
    2.2.3 Future environmental legislation.......... 50
2.3 Legislation on the working environment

3. VOLUNTARY AGREEMENTS BETWEEN THE INDUSTRIAL ACTORS................................ 53
3.1 Introduction
3.2 Agreements at National level
3.3 Agreements at Sectoral level..................... 54
3.4 Agreements at Company Level..................... 57
3.5 Agreements at Local Level....................... 58

4. POLICY STATEMENTS, DEMANDS AND CAMPAIGNS......... 60
4.1 Introduction
4.2 Management and Employer Strategies
4.3 Trade union strategies........................... 62
4.4 Green Movement Involvement in Campaigns.......... 64

5. CURRENT ENVIRONMENTAL CONFLICTS AND THE INDUSTRIAL ACTORS................................ 65

6. SUMMARY AND RECOMMENDATIONS..................... 67

   BIBLIOGRAPHY....................................... 71

1. INTRODUCTION

- The Development of Environmental Issues and Industrial Relations Environmental issues first made their appearance on the industrial relations scene in the mid 1960's as a result of workers struggles over health and safety in the Farmitalia pharmaceuticals plant at Settimo Torinese and the Solvay chemical plant at Rosignano. Three significant periods can be identified in analysing the features and results of trade union health and safety initiatives, the first from 1965 to 1975, the second from 1976 to 1985, and the third from 1986 to 1990. The characteristics of the different periods are directly linked to the economic cycle and to the nature of industrial relations.

First period (1965-75)

The first period saw the issue of health and safety in factories at the centre of claims and bargaining and policy initiatives taken by trade unions and workers at all levels, from works councils to national trade union organisations. During those years there was an explosion in company level bargaining, the works council emerged and the elimination of monetary compensation for dirty and dangerous work became a workers' objective. Health and safety was the subject of discussion and analysis and, in more pratical terms, of negotiating practice. Productive relations were established with the scientific community and universities resulting in increased interest in health issues.

The collective agreement covering workers in companies part-owned by the government which was signed in 1967 set up "committees for accident prevention and safety". This together with the development of company level bargaining gave rise to a series of discussions and agreements with the aim of improving working conditions, workers' health and safety and the elimination of compensation for dirty and dangerous work in short, the reduction and control of working environment hazards.

The last significant result of workers' initiatives, during this first period, was the agreement reached in 1974 by the trade unions and the petrochemical companies on the control of the production cycle of vinyl chloride monomer (VCM) which began at the Ravenna chemical plant. The trade unions had undertaken research into the health of workers exposed to VCM hazards, the results of which proved that this product was carcinogenic. It was therefore agreed with the producer companies that all Italian factories should change over to a closed cycle system of production.

Second period (1976-85)

This period saw a decline in interest in work environment questions, a decline which had both objective and subjective causes. The first cause was the economic crisis which affected the Italian production system, placing the question of jobs and employment at the forefront. The second cause was restructuring involving new process and product technologies and resulting in flexibility of plants, lack of maintenance and heavy use of plants on a continuous cycle basis. The works councils and trade unions were not adequately involved in this. The decline in attention and negotiating practice devoted to "working environment hazards" became dramatically prominent.

The change in the economic cycle, and the threat of job losses put the trade unions on the defensive. During the period of extreme restructuring a change in the industrial relations system, moving from company level negotiations to a centralised bargaining system, occured. In this system, a centralised structure of collective bargaining was established with a parallel co-ordinating structure for the various bargaining levels. Centralised bargaining took on the dual role of controlling the agreement structure and linking the social partners and the State1. Macro-economic bargaining became an established feature of industrial relations. Whatever evaluation is make, it became the main industrial relations tool to control growing inflation and stabilise conflict situations.

Third period (1986-90)

In the 1980's centralisation was replaced once again by decentralisation in many aspects of social and economic life, and not only in collective bargaining. The drive towards decentralised bargaining, encouraged by technological change, came mostly from the employers who for the first time took the initiative in industrial relations. Decentralisation went hand in hand with a tendency towards a fragmentation of collective bargaining, a reopening of pay differentials and a differentiation of industrial relations models. More recently, "deregulation" has become a dominant theme, used controversially against the rigidities of a regulatory approach, not only in industrial relations, but also in social life.

All these phenomena are aimed at greater flexibility in industrial relations. A growing need for flexibility is expressed by companies under pressure from the changed market conditions of their product including growing variability and unpredictability of demand, high interest rates and intense international competition. Flexibility in all its aspects is greatly favoured by new technology and consequent innovations in organisation and production. Many of these trends appear to some people to reduce the opportunity for "consensus seeking" as an instrument for regulating and stabilising industrial relations. Others feel that these same trends increase the opportunity for stable industrial relations.

Returning to the environment, this third period is characterised by a shift and expansion from "working environment hazards" to "environmental hazards". At the end of the 1970's the decline of trade union interest in environmental and health problems met with new and tumultuous feelings within Italian society. Industrial relations, and especially the initiatives taken by the trade unions had come to a halt at the factory boundary, unable to identify negotiating instruments suitable to control the impact on the surrounding area and on the natural resources. But society had moved on. Bhopal, Chernobyl and the referendum on nuclear power all provoked an explosion of ecological and environmental organisations and movements, finally reaching the trade unions and the employers' organisations.

At first, the connection between environmental conditions inside and outside the factory was not clear, and it was too early to identify positive research and innovations stimulated by environmental concerns. But the collective agreements of 1986 clearly opened up the issue of "environmental hazards" and produced controls covering the relationship between the factory and its surrounding area, the use of resources and external pollution. With these agreements the social partners again faced the much more complex and difficult concept of "the environment".

Today none of the social partners can deny the importance of environmental issues. This is not only because there is pressure from the mass media and green movement, but also because any management failing to include an environmental component in its policies is disregarding the relationship between the company and its image and between the product and its market. Environmental balance has also become an objective for both management and trade unions.

The crux of the matter is that the environment cannot be an additional objective but must be incorporated into all employer and trade union policies. There has been speculation, particularly within the unions, that "green industries", such as that manufacturing pollution control equipment, will mean an increase in job opportunities. But, it is likely that rather than produce additional employment it will be used to employ workers rendered "mobile" because of the suspension or the closure of polluting industries.

2. THE LEGAL FRAMEWORK

2.1 Introduction

The Italian legal system does not directly address the issue of industrial relations through regulations governing the environment and health and safety. There are unwritten practices and contractual agreements between the parties, including the state administration, which give new scope to the system of industrial relations including the environmental sector. A brief overview of certain laws concerning the environment and health and safety follows in this section.

2.2 Environmental legislation

In 1986 the first piece of environmental legislation containing legal rights was introduced, and this year can therefore be considered as the year which introduced the "right to the environment". Prior to 1986 environmental protection legislation did exist, but law no.349 setting up the Environment Ministry gave formal recognition to this right to all citizens, in particular the right to information on the state of the environment and any changes in it. Although the Environment Ministry came into being in 1986, it was from 1988 on that a real body of laws on environmental protection began to be enacted. The most important EC Directives on the environment number one hundred and twenty nine, eighty nine of which have been incorporated into national law and seventy seven of which are fully in force.

However, environmental legislation in Italy is basically in line with EC legislation, from Directive 501 on "major hazards" known as the Seveso Directive, to those establishing threshold limit values for sulphur dioxide, lead and so on to that requiring an environmental impact assessment in the case of major developments.

To return to the main topic of analysis, industrial relations, it should be pointed out that only two of the laws mentioned identify or indicate "procedures for the practice of industrial relations" and set out the role of the main actors, employers and trade unions. Article 12 of Presidential Decree no.175 (the Seveso Directive) states that information given to workers and to the trade union movement is of value in determining the "hazard index for major hazards". Article 7 of law no.349 regulates "areas with a high risk of environmental crises" and provides for the establishment of state and regions committees whose members also include, as an experiment, employers' and trade union organisations. These are examined in more detail in the following sections.

2.2.1 Areas with a high risk of environmental crisis

The legal entity of "a high risk area" came into being with the law establishing the Environment Ministry. Article 7 of that law states that those areas "characterised by serious alterations of their ecological balance as regards water, air or land, are to be declared -by the Council of Ministers (the Italian Cabinet) on a proposal by the Environment Ministry, in agreement with the regions concerned - to be areas with a high risk of environmental crisis".

In this declaration, the objectives of reclamation are identified. The pollution recovery and reclamation plan details the action necessary action to restore environmental balance. If the region or regions involved fail to implement the plan, the Environment Ministry is to intervene and must take direct action. The features of this environmental policy instrument are:

(i) a shift to central government, under the Environment Ministry in consultation with the regions, of the power to take action in these "crisis areas". In Italy, environment policies are normally under the control of regions or communes, which often fail to exercise their regulatory functions. In this case, the Environment Ministry is responsible for coordination and planning action programmes, but also, when necessary, can completely take over as the responsible authority;

(ii) the establishment of the state and regions committees, an important coordinating body which can include, apart from Environment Ministry representatives who presides over it, the minister of health, industry, agriculture or marine depending on the topic being discussed; the regions, provinces and communes (the local administrations) involved in the "risk area"; experts from the state technical services or other technical and scientific institutes; representatives of companies present in the area; trade union representatives and works councils involved and representatives of environmental organisations at local and national level. Indeed all the actors involved in the environmental dispute.

This new committee for handling environmental disputes has led to some significant results in two of the most difficult cases in Italy, the Valle Bormida involving the Acna chemical company in Cengio; and the chemical company Enichem plant at Manfredonia.

Environmental disputes often arise outside the factories and involve various social actors. These disputes need to be resolved in a neutral way, almost by arbitration, in order to take account of all the different views and the state and regions committee is a suitable body for this. It encourages agreement on reclamation plans, but also leaves the parties free to engage in unilateral actions of commitment, letters of intent, agreement protocols or even to reach simple verbal agreements. The philosophy is to aim for a result even if this involves freely entered into unilateral agreements. It is no longer always an exercise imposed from above but rather a joint search for a compromise which can be accepted and achieved in agreed time limits.

2.2.2   Draft agreement between the Environment Ministry and the trade union federations, CGIL, CISL and UIL.

Environmental policy has reached an important turning point where either a basis is laid for a move away from emergency mesures to real planning, or environmental policy in Italy will become deadlocked and totally out of line with the new European Context of 1993. The Environment Ministry and trade union organisations agreed on the need to identify operational instruments, resources, decision-making processes and administrative channels suitable for an environmental protection policy, beginning with a 1992 Financial Act.

In order to promote such a policy the parties agreed that:
the trade union organisations will meet with the environment minister to discuss the drafting of a new three year environmental programme from 1992 to 1994 giving special consideration to all aspects involving environmental reconversion policies, the impact on employment and innovative processes; within the framework of the new Committee of the State Departments (CIPE) decision implementing the three year programme, provision will be made for allowing trade union organisations to submit projects relating to research, monitoring, training, education and new jobs; the environment minister recognises the need expressed by the trade union organisations to fully participate in the state and regions committees of high environmental risk areas. The minster will promote an experiment which, based on area by area monitoring, will gradually ensure the presence of the trade union movement on such bodies. Initially priority will be given to those areas where the environmental crisis is most linked to industrial activities.

Furthermore, in view of the publication of a map of areas with the highest concentration of industries posing a risk of major industrial accident by the Ministry (pursuant to Presidential Decree 175/88), it was agreed that the trade union organisations CGIL, CISL and UIL would join the committees which are to operate in such areas.

Agreement has been reached on a number of areas concerning environmental programme agreements which the minister has signed or is about to sign with major public and private groups. These include consultation with the minister with regard to the trade unions' programme agreements, highlighting of all aspects of the programme which may involve negotiations with trade unions, and that negotiations will involve the Environment Ministry, the Labour Ministry and the various companies concerned.

2.2.3 Future environmental legislation

The draft "eco-audit" directive and the comments submitted by the CES at a meeting on 12 June 1991 with the management of DG.XI (the directorate general of the European Community with responsibility for environmental issues) have resulted in a great deal of interest in this new area of industrial relations. The directive is concerned with environmental impact and the use of resources. It provides for "company committees" on the environment; access to environmental information; audit of results achieved through environmental mesures; the promotion of harmonised standards for the management of environmental resources at firm level and an annual presentation of a "declaration on the environnement" by plant managers (The Draft Regulation including the observations of the ETUC are enclosed).

2.3 Legislation on the working environment

If environmental questions are considered as being inextricable from the general problem of protecting the health of workers and citizens, then there are starting points in legislation. Article 32 of the Italian Constitution declares, "The Republic protects health as a fundamental right of the individual and an interest of the community." No consideration is given to the protection of environment, although article 9, speaking of culture and scientific research, specifies that "The Republic protects the countryside and the nation's historical and artistic heritage."

Similarly the Workers' Rights Bill, which is the fundamental law regulating labour and union relationships, ignores environmental questions while strenuously reaffirming the protection of the physical integrity and health of workers. Article 9 states that "Workers, through their representatives, are entitled to monitor compliance with standards to prevent accidents and occupational diseases and to promote the study, preparation and implementation of all measures necessary to protect their health and physical integrity".

Law no. 833 of 1978, known as the Health Reform Act, assigned the power of inspection, control, and health protection and prevention in factories to local health areas (Unità Sanitarie Locali - USL), referring to the more general protection of the external environment, the living environment of workers and citizens. The control of workers' health and safety and the quality of citizens' life in Italy is still assigned to the local health areas. National laws governing health in the workplace and the prevention of occupational accidents date back to the 1950's (DPR no.303, 19 March 1956 and DPR no.549,

27 April 1955). On the 17 August 1991 the government issued a new legislative decree implementing European Community (EC) directives pertaining to the protection of workers from exposure to chemical, physical and biological agents in the workplace. Article 5 of this decree provides for "a discussion of information relating to specific risks caused by exposure to chemical, physical or biological agents, as well as control by workers and their representatives on the application of protection measures".

The entire issue of workers' health and safety is undergoing legislative review in Italy. Parliament has delegated the government to implement all EC directives relating to health and safety, including the so called "framework directive" which is of much interest to the industrial relations actors. At the same time, a parliamentary commission is consolidating the legislation into a single Act. It is interesting to note that this parliamentary commission, known as Lama commission after the name of the Senator who is chairing it, has already drafted an initial bill which provides for the appointment of a "safety delegate".

The safety delegate would be appointed as a representative of the workforce to control health and safety standards and to promote, together with trade unions, public services and employers research and development of any solution necessary to improve working conditions in the workplace. Under the proposals, the safety delegate is entitled: to verify and monitor compliance with health and safety standards in the workplace, as well as environmental regulations relating to wastes, effluents and emissions and the handling and storage of hazardous substances and preparations; to inform and warn workers, corporate agents, prevention services, trade union representatives and managers, of any risk identified; to propose, during periodic meetings, all necessary preventative measures; to express an opinion on the prevention plan which the employer (management) is required to prepare within three months from the enactment of the legislation. A training plan constitutes an integral part of the prevention plan.

In case of non-fulfilment of these regulations the safety delegate must inform the competent authority. In case of any imminent or serious danger the safety delegate must promptly inform workers who have the right to refuse dangerous work: "Every worker has the right to suspend his work in the presence of the risk caused by the violation of health and safety standards in the workplace, whether or not there is a violation of the legal regulations"

Even the early EC programmes of action on health and safety tended to promote the participation of the social partners in decision-making at all levels, in particular at plant level. Now more than ever, EC policy sees the participation of social partners as an important, indeed essential, tool in the planning and implementation of any initiative designed to improve working and living conditions.

This policy is developed and given effect in Directive 391/89, the framework directive, which singles out (article 11) the "consultation and balanced participation of the social partners" as the hinge of Community and member States policies. The directive contains a number of requirements relevent to industrial relations ranging from the recognition of "workers' representatives with specific functions in the area of health and safety" to provision of information, preparation of training and prevention courses and consultation (article 6) of social partners as well as the planned introduction of new technologies and their impact on the organisation of work (Directive 391/89 enclosed).

3. VOLUNTARY AGREEMENTS BETWEEN THE INDUSTRIAL ACTORS

3.1 Introduction

In Italy the voluntary arrangements, namely national collective agreements, company agreements, territorial agreements and those negotiated at the workplace, are the most important source of environmental industrial relations. Direct negotiation between employers and workers at branch, company and workplace levels, have established rights and obligations for both parties on matters concerning the environment. These agreements shape the new models for industrial relations based on "confrontation" and "joint decision making". The analysis that follows takes account of national collective agreements in the manufacturing sector, agreements between the Employers Confederation (Confindustria) and trade unions, agreements signed by large manufacturing groups and territorial agreements.

3.2 Agreements at National level

In June 1991 negotiations between Confindustria, the trade union federations and the government began, aimed at redefining the wages structure, the "bargaining model" and other mechanisms ruling certain contract matters, for example the "sliding-scale" wages index mechanism. The environment was included among the topics selected for discussion.

Negotiations were based on the theory that in order to solve environmental problems, the following are necessary: placing environmental requirements in the economic and social fields of the social development process; defeating the "culture of catastrophe" i.e. deindustrialisation and recession; considering recourse to "conflict" as a residual and extreme form with regard to environmental policy interventions and confirmation that solutions to environmental problems depend upon a correct technical know how and certainty and applicability of standards.

This would involve the expansion of the industrial relations system to allow the joint evaluation of environmental matters; the "time, ways and means" for initiatives, including joint initiatives, to be taken at every level of the state administration and the proposal of solutions that are in line with the processes of the economic and social development of the country.

An organisation known as an environment observatory has been proposed which would manage confrontation between unions and employers in areas such as environmental legislation, with possible proposals being submitted to Parliament and government; develop projects and suitable policy instruments, including financial ones, aimed at encouraging industrial reconversion, labour mobility and relocation; carry out and suggest research in the environmental field and undertake vocational training initiatives on safety and environmental issues.

In addition the observatory would give support to the three sided confrontation between governmment, unions and employers' organisations for planning environmental policy interventions which entails a great deal of work for those responsible for industrial relations within companies and trade unions. Although at national level agreement has not yet been reached between Confindustria and the trade union federations, at regional level in Lombardy, an important agreement was signed 10 July 1991 by the association of Lombard employers (Assolombarda- the most important in Italy) and the Lombard CGIL-CISL-UIL.

In this agreement the contracting parties acknowledge that health and safety and the environmental issues are matters of high-priority for all social parties involved. They agreed upon the setting up of a joint commission for the study of environmental problems with its headquarters at the Assolombarda. The commission has various functions, including encouraging initiatives in the field of vocational training on environment and safety at work and acting as a support body in the case of a legal dispute and facilitating conciliation between company management and union representatives. Such interventions must occur on a voluntary basis and at the joint request of the parties involved prior to appeal to the courts. The commission also carries out research on fundamental environmental problems both inside and outside the factory, and into possible reconversion in the manufacturing sector and the relocation of factories from one industrial area to another.

3.3 Agreements at sectoral level

At present the sectoral level in Italy is the basic level of collective bargaining, the real forum on almost all questions of industrial policies and the organisation of production and work and there are three levels. The first is the one which produces the industry-wide agreement and the actors involved are the trade unions federations representing the occupational workers' groups on one side, and the employers' federations representing the sector on the other side. For example, in the chemical sector, there is the FULC (Federazione unitaria lavoratori chimici, united federation of chemical workers) on one side, and on the other the Federchimica-ASAP (Associazioni degli imprenditori chimici privati e pubblici, Associations of private and public chemical employers).

After the collective agreement, which can remain in force for three or four years, has been signed there is a second level of bargaining, the so-called "group accord" and finally a third one at workplace level. The second and third levels form a part of the decentralised bargaining process, which for some years has tended to emphasise the plant rather than the industrial group.

The sectoral level, in particular the industrial-wide agreements, are highly important tools for negotiating procedures, relation models and content of environmental questions. The collective agreement between ENI and the chemical and energy workers' union back in 1967 provided for "prevention and safety committees" designed to guarantee health in the workplace.

Today the collective agreement for the chemical sector is the most advanced as far as environmental industrial relations are concerned. From an examination of national agreements currently in force, it appears that those signed in the metal, textile and other sectors, although providing for health and safety in the workplace, only marginally deal with purely environmental questions and fail to provide for "ad hoc" relations and procedures.

The industry-wide agreement for the chemical sector promotes industrial relations models open to the possibility of joint responsibility and decision-making. The most important points of the environment chapter, article 42, of this agreement (enclosed) are summarised below.

This article formally recognises environmental representatives, workers representatives responsible for bargaining on environmental issues, the committee for health protection in the working environment and the environment commission, formed by environmental representatives from different production areas. A training course of a minimum of 150 hours for environmental representatives at the firm's expense is provided for.

Article 12 reconfirms all the rights and powers already outlined in previous contracts and attributed to CdF (Consigli di Fabbrica, works councils). Discussions with company management are provided for on investments aimed at improving environmental and ecological conditions; training programmes designed to improve safety; topics concerned with atmospheric emissions, discharges into water ways and solid wastes; regulatory initiatives at national or EC level concerned with safety, health and the environment; problems relating to carcinogenic or mutagenic substances; plant safety files; safety reports which must be submitted to the Environment Ministry and technical solutions and innovations to prevent or control risks.

There are two interesting points in Article 12. If management wishes to introduce new substances in the production cycle, the use of which could give rise to new health or environmental risks, or alter the production cycle by new technologies and possible new risks, this should form the subject matter of a "prevention dispute" between the works council and management. And in companies with more than 300 employees, an environment programme must be presented at a special meeting. This must contain management goals with regard to environmental improvements made to products, technologies and infrastructures and how these improvements will influence the environment and health and safety conditions both inside and outside the plant.

Article 42 paragraph 5 provides for the examination by management and the works council of recovery and or restructuring programmes for environmental and safety reasons involving major changes to plant, or the total or partial cessation of the plant with job losses. During the examination of the recovery programme, which could last 20 days at the most from notification from the company, unions and employers should not unilaterally take any step resulting in dispute.

Article 43 of the agreement governs "prevention, hygiene and safety at work" and includes new procedures and puts a greater emphasis on environmental industrial relations especially as regards information and the safety training of workers (safety index card for hazardous substances used, index card for plant characteristics).

The most innovative part of the agreement is the chapter governing "Industrial Relations" and the provision for the establishment of a national observatory. The agreement says,

"The association of chemical employers (Federchimici-Asap) and the chemical workers union FULC, aware of the important role played by industrial relations in contributing to the solution of economic and social problems and guiding the action of its representatives, in the light of past experiences, call for the constitution of a "national observatory" in order to provide their own contributions and proposals for the formulation of trends in the safety and environment field".

The observatory will be made up of representatives from trade unions and employers and will have the following responsibilities,

a) comparing each other's attitudes in relation to the development of national and EC standards on environmental issues, and selecting possible proposals to be submitted to the competent authorities;

b) carrying out joint evaluation of initiatives in environmental and safety areas;

c) monitoring the development of environmental and safety conditions in the sector, taking account of fundamental problems connected with product technology, plant relocation or recovery programmes;

d) identifying common proposals to facilitate the management of legal obligations and methods of interaction with enforcment agencies;

e) identifying content and structure to promote environmental and safety training, with particular reference to company appointed technicians and members of the committe for

health protection in the working environment;

f) dealing with topics concerning atmospheric emissions, liquid and solid wastes on the basis of available knowledge and

g) examining problems relating to carcinogenic or mutagenic substances.

3.4 Agreements at Company Level

Two typical agreements signed are the agreements reached between the domestic appliance manufacturing company Zanussi (owned by the Swedish multinational, Electrolux) and the Federazione dei lavoratori metalmeccanici, FLM (Federation of metal and mechanical workers) covering 15 000 workers; and between Enichem, an Italian chemical company owned by Ente Nazionale Idrocarburi (hence "a state holding" but well known also abroad) and the FULC covering around 40 000 workers.

The Zanussi agreement follows the co-decision making industrial relations model. The agreement provides for the setting up of joint committees with decision-making power on a number of issues, such as environment, vocational training, technological innnovations, company canteens and so on.

"A dispute between the parties", said Luciano Scalia, the national secretary of chemical union FIM-CISL, "will always be prevented, because a decision has to be reached within the committees after a thorough discussion".

The Enichem-FULC agreement is based on the theory that the environment may become a theme around which to experiment shared models of responsability involving the industrial relations actors, employers and workers. In order to encourage openess and increase the flow of information, the parties agreed to set up a joint committee able to examine environmental situations. In carrying out its duties, the committee will review and deal with problems posed by the existing technical production organisation in factories, seeking to identify solutions necessary to improve environmental compatibility. This joint examination should allow a moving away from an environmental policy based on "end-of-pipe" interventions, such as waste treatment and dust arrestment, to an innovative policy focused on "clean technology".

The agreement also provides for access to information and environmental training aimed at enhancing workers' technical and specialist knowledge regarding the environment and safety in order to increase awareness about environmental issues. The chapter on training concludes that it is essential to further strengthen the level of participation and involvement of those concerned in the production process.

Specific commitments in the agreement concern the replacement of highly polluting mercury cells in the production of "clorosoda" with "membrane cells" which are more environmentally compatible; and the reduction of carbon dioxide and sulphur dioxide emissions; completion of the waste water purification system; reduction in water consumption through new cooling plants; monitoring the waste disposal system and carrying out environmentally orientated research and development.

3.5 Agreements at Local Level

The industrial actors have been involved in agreements concerning the local environment involving area planning, the rational management of waters within a hydrographic basin, the transport system, the management of discharges and waste disposal, the reclamation and rehabilitation of former industrial sites and area, the decentralisation of activities, the use of reclaimed and rehabilitated areas, the quality of air or water within a district or city and areas set aside for public green spaces and parks.

The agreement concerning the resiting and relocation of several petrochemical plants in the Genoa region (involving oil and chemical companies I.P., SAAR, ESSO, PIR Carmagnani and Superba) was signed not only by the unions in the area, the employer associations and chemical companies, but also by the Genoa City Council and the Genoa Port Authority.

The agreement concerning the improvement of air and water quality in the Val Chiavenna valley in Lombardy included the improvement of atmospheric emissions from several plants owned by the metallurgical company Falck, changes in the methods of disposal of industrial wastes and effluent, monitoring systems, epidemiological studies of the population in the Val Chiavenna in relation to the effects of chromium emissions, and presentation of a general plan for the environmental reclamation of the Val Chiavenna district. It was signed not only by Falck, the works council and trade union organisations, but also by representatives of the mountain communities, the mayors of the various villages in the valley, environment movement representatives and the Environment Ministry.

The "Utopia" project provides for the relocation of most steel works in Italy from Genoa and Naples to more appropriate areas for environmental reasons. This project, presented to the parties by the minister for economic planning, was examined and discussed until a preliminary agreement was reached which led to a " programme agreement" (enclosed). This was signed by the Italian government (minister for the environment and economic planning of urban areas), the presidents of the regions concerned, Liguria, Tuscany and Campania, the mayors of the municipalities concerned, CGIL-CISL-UIL trade unions, IRI, the most powerful public economic institute and ILVA, the iron and steel firm concerned.

The programme agreement has become one of the instruments for joint decision making on environmental programmes. Other developments are still undergoing discussion, in particular reclamation of the Po river and the Adriatic sea, which has been affected by algal growth as a result of excess nutrients (eutrophication), defence from high tides and reclamaion of the Venice lagoon and improvement of air quality in Milan.

At this local level of negotiation new actors are involved who although are only informally provided for within the industrial relations system, play a full role of decisive importance in dealing with environmental questions which concern them. These new actors include environmental organisations, communities, local bodies, and industrial unions.

4. POLICY STATEMENTS, DEMANDS AND CAMPAIGNS

4.1 Introduction

Today no citizen, whatever their social class or educational background, would feel able to deny that environmental problems exist. But not all agree on the action that must be taken in order to make development compatible and sustainable. Some blame technological development and hope for a return to pre-industrial society. Others believe that it is too late to do anything and wait for the catastrophe. A third group believes that the solution to environmental problems cannot be attained by rejecting science and technology, but rather through a different kind of development.

Workers and employers, as principal actors in the industrial relations scene, have for some years seen society gradually recognising the central importance of environmental questions, but their attitude was one of "wait and see", considering the issues as marginal to the industrial relations system. In Italy this waiting time was longer than in the other Community countries. The disaster at the Icmesa chemical plant in Seveso, the explosions in chemical plants at Manfredonia and Priolo, the accident in the chemical Farmoplant at Massa Carrara and the eutrophication of the Adriatic, were not enough.

In short the real pressure has come from outside, from the green movement, the press and politically progressive scientists and magistrates. Only recently has remedial action been taken with new rules being developed within the industrial relations system allowing dialogue not only on questions concerning workers' health but also environment protection.

4.2 Management and Employer Strategies

Following a period of uncertainty, it is now several years since the Association of Italian Employers made a significant commitment to environmental topics, promoting a whole series of activities ranging from the establishment of an industrial-environmental association and an environment institute, to the publication of a review called "Impresa-Ambiente", to arranging a series of seminars and meetings on environmental questions.

The president of Confindustria, Sergio Pininfarina, stating the ecological commitment of Italian firms declared, "the protection of the environment is not only compatible with economic development, but the ecological problem, if correctly tackled, can indeed be a powerful factor for development and improvement of society. Industry has the necessary technological know-how to adopt suitable solutions for removing pollution from land, air and water." 2

Confindustria calls on the state and public authorities for discussion. It maintains that the main objective must be to provide incentives for less polluting production processes without distorting competition or hindering fiscal harmonisation. It claims the right to participate in the preparation of environmental policy in Italy, given that companies are active protagonists in the environmental policies of the country, due to the obligations and responsibilities which fall on them.

Confindustria has prepared six themes on environmental commitment. These are compatibility between development and environment; compatibility between technology and nature (confidence in human abilities); consistency of decision taking in uncertain situations; the link between the level of environmental protection and costs (individual and collective); the environment as a global problem (international cooperation), and the circular nature of the problem (relationship between the internal and external environment of the workplace).

As regards individual companies, some of them are more aware of environment issues than others. But there is no doubt that "anyone who has not yet understood the central importance of this theme will sooner or later be forced to do so, because it is going to be increasingly difficult to run a business without providing a managerial and strategic structure able to tackle environmental issues" (Carlo M. Guerci, Professor of Political Economy - Genoa Impresa - Ambiente, no 1, 1990).

This is true for two reasons. The first is that environmental legislation will inevitably increase and the second is that the employer will also have to cope with both the company and the consumer market which will increasingly reward companies which produce goods compatible with a more protected environment and whose production process is ecologically clean. The spread of a proper environmental culture will have a growing impact on companies in terms of collective consumer demand. The products which receive most attention from an ecological point of view will be those most highly rewarded by consumers in the marketplace. This means that even those companies which today play a passive role in relation to the environment will have to face the problem and to play "a self-regulating role" in promoting environmental issues.

Some large companies in Italy, have already done this. For example, chemical company Ferruzzi-Montedison has launched a new biodegradable plastic material made of natural ingredients and ENI produces "green" petrol with a low lead content. These two companies have set up environmental departments and work to create an "environmental culture" through training courses. These are very large companies whose operations have a high risk of environmental hazard and which offer opportunities for trying new ideas in industrial relations and the environment.

The trade union organisations and works councils are trying out joint decision processes with them. These have not yet reached the "right to participation" stage, but the "resistances" to developing a more courageous role of industrial relations are beginning to give way to a new attitude of "willingness" in these companies.

4.3 Trade union strategies

For the three big trade union organisations in Italy, CGIL, CISL and UIL, the biennium 1989-1990 was dedicated to the environment. National meetings, programmes and public initiatives were undertaken to express the commitment on environmental questions from health protection in the factory to the wider protection of the environment, from working hazards to environmental hazards.

At the end of the 1980's the referendum on nuclear energy saw a large part of the workers' movement take a position against this type of energy, but there was no move to a wider vision of environmental balance. Today the environment is part of the consciousness of many union leaders, and has been officially ratified in the statements by Franco Marini, the General Secretary of CISL during the May Day celebrations (1989, May, 1 - Venice), by Giorgio Benvenuto of the UIL, in "Verde UIL", and in the report presented by Bruno Trentin, General secretary of CGIL, at the programme conference held in Chianciano in 1989.

According to Trentin, the "quantitative development of the economy, the growth in the production of goods and services, and the development of employment, are increasingly clashing with objective structural limitations, of which our collective culture is becoming gradually aware; they must deal with ever closer constraints. What are these limitations and these new constraints? First of all, the limitation represented by the possible destruction of the ecological balance in the world. Reconciling development with health, the biological progress of people, subjecting this progress to the constraint of a different relationship with nature, guaranteeing its survival and growth, means taking on ourselves, as trade unionists, the objective of managing development in such a way that environment, health, ecology and culture can be created as well as goods. It means rethinking our idea of development in terms of energy saving, in terms of product duration, in terms of product quality".

This statement of principle is matched with a full-scale programme of work from the environment department of the CGIL (the CISL and UIL also have environmental programmes). The different aspects of this programme which range from the greenhouse effect and CFCs to a re-examination of the debts of developing countries (the relationship between North and South in the world) with a direct and immediate impact on industrial relations are listed below.

A draft proposal for reconverting production facilities so as to make them environmentally consistent; handling the implementation of the "Seveso Directive" in the workplaces involving the compilation of safety cards, safety reports and procedures for workers' information and training; the application of procedures for evaluating the environmental impact of new investment (the environmental impact assesment); making contact with the government to encourage scientific research for environmental purposes, capable of producing major innovations in production plants; environmental controls, environmental registers, environmental budgets covering plants and industrial groups; regional disputes (the Adriatic dispute, the Venice dispute); agreements and contracts at national and company level; application of EC environmental policies and harmonisation in a view of 1992 and the Single Market; use of economic instruments for environmental purposes; integrated management of resourses in major hydrographic basin and reform of the Environment Ministry.

Trade union organisations have also developed relations with environmental organisations which often play fully active roles in negotiating environmental questions. Joint seminars and informal meetings have been organised with environmentalists, in order to debate areas such as proposals for legislation.

There are also environmental organisations, including one called "Ambiente e Lavoro" (Environment and Labour), which are very close to the trade union organisations, in fact almost an offshoot of them. Its headquarters are in Milan, and as well as performing a significant service function for union branches and works councils, it promotes initiatives and meetings involving employers' associations, the Environment Ministry, specialists working on accidents prevention and the protection of health and the environment, and the trade union movement itself.

Rather less impressive is the area of "emergency management". Accidents, contravention of standards and water and air pollution require the unions, employers, green movement and the public authorities at different levels to meet and agree positions very rapidly to deal with these emergencies. The lack of legislation and the threat of job losses means that workers and trade unionists at local level often adopt positions which are not fully consistent with their own union's programmes and environmental commitments. Italian unions are in the early stages of environmental policy and action and inconsistencies are to be expected as they try to overcome the "emergency" point of view which results in a choice being made between employment and environmental protection.

4.4 Green Movement Involvement in Campaigns

In Italy, the Green Party and environmental movement, as well as demanding more environment-friendly production processes, have advanced complete proposals for reconversion in the plastics and agricultural chemicals sectors. In some regions laws have been passed granting incentives to alternative production, for example biological agriculture.

One line adopted by the environmentalists is participation in shareholder meetings through the "green shareholder movement", demanding major modifications in processes or products. This has been done in Montedison, Fiat and Sip, for example.

Company managements and trade unions have been willing to undertake informal and formal arrangements with the environmentalists, researchers and the scientific world in order to implement, where possible, environmental reconversion. Environmental organisations are often involved in environment negotiations. Trade union organisations often act together with them in embarking upon new initiatives, such as the holding of joint seminars and formal and informal meetings to air opinions on, for example, new legislation. Recently trade union organisations and the most important environmental organisations, the Lega Ambiente (Environment League) and the Friends of the Earth, have drawn up an interesting "draft bill for reconverting productive activity for environmental purposes".

This, officially presented to the press, all parties involved and parliamentary groupings, calls for the creation of a fund for financing reconversions, a technical body under public control to assess the state of environmental crisis and prepare a recovery plan and a "green" wages supplement fund for those workers temporarily unemployed while awaiting re-employment in environmental friendly production activities.

But there are also cases, for example Acna and Enichem of Mafredonia, where relations between trade union organisations, employers and the green movement are not good, indeed almost violent at local level. Until recently collaboration, even where there is a total clash of views, has always been salvaged at national level with the assistance of associations of national interest, linked to the green movement.

5. CURRENT ENVIRONMENTAL CONFLICTS AND THE INDUSTRIAL ACTORS

The attitude and behaviour of employers and trade unions facing important environmental conflicts is difficult to summarise. Attitudes are sometimes contradictory and require a deeper complex analysis. For example, both trade unions and employers' organisations agree that it is necessary and urgent to restrict economic growth and enter the culture of sustainable development. But is is almost impossible to set limits to some processes, such as cars and pesticide production, construction of buildings and new roads, or the use of plastics. In several public initiatives employers and workers have discussed industrial reconversion but this has not been taken seriously.

Results have been achieved in the nuclear field and are continuing. Following a referendum that saw the trade unions sided against nuclear energy for civil and military use, the industry has been completely reconverted so that nuclear power stations now work with natural gas in Italy. In 1991 the debate concerning the usefulness of nuclear research was reopened. And with regard to the greenhouse effect and ozone depletion and the elimination of fluorocarbons and control of carbon dioxide and sulphur dioxide emissions, the Environment Ministry launched a programme which gained the consent of the chemical and energy industry and trade unions. But initiatives aimed at controlling and reducing the noxious emissions of cars exaust, or closing city centres to traffic have been hindered.

A number of large companies such as ENI (Ente nazionale Idrocarburi - the national oil corporation) appear to be seriously committed to the study of North-South global relations and in particular the problem of protecting the Amazon region in view of the United Nations conference in Rio de Janeiro (1992) on environmental problems (1).

While there have been some successes which suggest that differences and contradictions between statements of principle and real behaviour do not exist or are very limited, in reality there are some areas, for example water conservation and waste disposal, which constitute "areas of contradiction" especially with reference to the conduct of small businesses. The disposal of industrial waste in Italy is the most critical area where there are many cases of illegal practice.

Relocation and recovery plans tend to be accepted by employers but the problem of who pays the costs has not been resolved. The shutting down of incompatible industrial plants gives rise to strong resistance both by trade unionists and employers. A striking example is that of the chemical company Farmoplant in Massa Carrara. Until the last moment both the trade unions and the company tried to save the plant from shut-down despite a very serious accident. Job losses without alternative employment is a very serious problem that cannot be underestimated or ignored. In the end, the Farmoplant shut-down was unavoidable as public opinion and the attitude of the local authorities left no alternative.

In order to conclude this chapter some recent experiences, both negative and positive where the parties involved behaved consistently with their official statements are related.

On the negative side Isochimica, an asbestos removal company in Avellino, was operating without providing protection from exposure for its workforce and disposing of the asbestos in an unlicensed site. And Balangero mine workers, in Piedmont, continue to defend this mining activity, objecting to the shut-down despite the ascertained risks incurred by workers and the nearby population.

On the positive side, in the agricultural sector, FLAI (the agricultural workers' federation) demanded the elimination of a number of pesticides and a more rational use of some others, to minimise food residues. An interesting consultation process began between the trade union movement and Confagricoltura(the agricultural employers' association), which may yield results which would have seemed unrealistic up to a short while ago. The metal workers' union and Fiat, in a discussion process chaired by the Environment Minister, reached an agreement to discontinue the use of asbestos in car brakes and clutches.

Employees working on insulation in large chemical and energy plants at Porto Marghera, Venice, were allowed by management to replace asbestos with synthetic fibres. And at Alfa Romeo, Milan the works cuncil, with technical assistance from outside researchers, drew up a proposal to replace solvent-based vehicle paints with water-based paints. The discussion process seems likely to produce good results, and it involved environmentalists and the public.

Employers, management and trade unionists are realising that the questions associated with the environment must form part of a specific branch of industrial relations and that they should and can be managed together. The Italian experience shows, however, that there is not yet a genuine awareness of this fact. Very often it is emergency management which makes management and workers aware of what is at stake. This does not give rise to a true "environmental compatibility culture". When the emergency is over the awareness too comes to an end and even the sharing of information is avoided. In some cases, although this is increasingly rare, the process of joint decision-making between management and works councils on these questions are sought only if they can be transformed into an alliance against public authorities, the public, environmentalists and the media. It should be pointed out that there is a new awareness by management and workers in large production units belonging to large public and private groups.

6. SUMMARY AND RECOMMENDATIONS

The review of state and contractual standards relating to environmental industrial relations allows a summary and identification of the innovations both taking place and which may take place in the "industrial relations system". In considering the hierachical order of laws, certain EC directives already or in the process of being issued are largely instrumental in broadening the scope of industrial relations to the environment, in terms of access to information and participation in decision-making (environmental committees); and the control and auditing of the extent to which the objectives set have been achieved (Declaration sur l'Environnement). Environmental legislation in Italy has a limited number of standards linking the Environment Ministry which provide for a "negotiated environmental model open to all the social partners affected and concerned by the environmental question up for discussion" as set out in the decree establishing the "areas with a high risk of environmental crisis" and the state and regions committees.

A closer examination of the role of the safety delegate will also be necessary if the bill is passed. This bill would require management to submit each year their prevention and training schemes and would give workers the right to refuse dangerous work.

The voluntary agreements signed by the partners are undoubtedly the most innovative and rich source of environmental industrial relations. These are the most certain, have been in force for a few years and have the advantage of being prepared directly by the social partners on the basis of the experience and facts which have seen the people concerned as real actors in the various situations. Therefore, special recognition must be given to the role played by voluntary agreements at different levels. They are particularly welcome and it must be pointed out that the most innovative ones concern mainly the chemical sector. Negotiation at national and industrial group level is the most prolific, and it is major companies which are mainly involved.

The most important innovations include the reconfirmation of the commission for health protection in the work places and the rights to information, monitoring and carrying out specific research and consultation in advance concerning the introduction of new substances and new technologies. Mention should be made of the innovative joint declaration by the social partners stating that a model of relations based on co-responsability and co-decision making with respect to the environmental question could be experimented with. It provides for the drawing up environmental programmes in factories with over 300 employees.

The national environmental observatory will become a forum for negotiation. The Enichem-FULC and Zamussi-FLM agreements make specific reference to "agreement hypothesis". Training in environmental matters has become an acquired right exercised by the social partners in several industrial sectors and small firms.

With regard to shortcomings and problems still to be solved, all the contracting parties fail to exercise these newly acquired rights on a continuous basis and it is necessary to further extend the new industrial relations model to all industrial sectors. There should be no difference in the way in which rights are exercised by major companies and medium and small size companies. There should be no question of workers and employers with preferential status complying with new industrial relations models while others find themselves excluded. To avoid this once the "new relations model" has completed the experimental phase, it could be extended to all activities, by passing legislation.

Environmental questions are now being addressed through a negotiation model which attaches priority to a "cooperative approach" involving the participation of social partners begins at decision-making level and not subsequently. Therefore the procedures governing the negotiation process are particulary important. A review of experience in Italy seems to reveal the presence of an "enlarged industrial relation model".

Three phases or spheres of action with respect to the enlarged model can be identified. A first phase where information, training and communication are the key words. Negotiation is and continues to be conducted between the trade union and the employer, but a level of joint responsibility (safety delegate, environmental commission, common research, training on environmental questions) is being provided for.

A second phase is where the negotiation process in many cases sees the participation of local or national experts and authorities and the relationship between the parties is formalised by joint committees, environmental observatories, environmental plans and audits of objectives as provided for by certain national contracts, group agreements and by the "eco audit" recommendation currently under discussion at the European Commission. During this second phase, decisions may be taken between two or three parties. This industrial relations model can elicit the broadest possible consensus because it does not lessen the influence of the traditional actors, the employers and trade unions, but allows contribution from other actors such as ministries or local authorities.

There is a third phase, which in some respects is to be welcome, which may take account of the concept of "enlarged environmental negotiations" and sees the participation of citizens' associations, environmental or consumers' organisations.

The "general model" could be represented graphically in the form of concentric circles (first, second, third phases) allowing for passage from one participation model to another according to the relevant problems and situations. The third phase or "the enlarged industrial relations model" is developing in Italy and may be indentified with the state-regions committee. An important suggestion may be to not destabilise the industrial relation already in existence but to proceed gradually. Each practical example should indicate the best type of negotiation to be experimented with. To force situations, to impose a more enlarged model on subjects and actors who are not ready for it could limit the scope of communication instead of broadening it.

In answer to the question as to whether in Italy the environmental question can became an issue around which joint decision-making may be created, employers and trade unions agree that environment is an important area to practice joint decision-making, but the practice is so far under developed.

The EC task force has prepared a report, "The environment and the consolidated market" which stresses that there is still a long way to be covered in order to harmonise environmental policies and laws in Europe and the major commitment required from the social partners to ensure that 1992 will not only promote the single market but also the environment.

Sustainable development can only be pursued through a process of reconversion which should take into account the reconversion of industrial zones or areas to achieve an overall reduction of pollution, the reconversion of the production process to optimise resources and reduce emissions and product reconversion to make goods more compatible with the environment and bring mass consumption mentality back to that of favouring quality to quantity.

It emerges from documents available that the trade union organisations will struggle to achieve all the requirements of EC Directives, national legislation and voluntary agreements concerning environmental matters, in particular the right of workers and citizens to environmental information and for new provisions to extend and certify such rights. The "instruments" which could be used to achieved such rights could be observers, joint committees, agreements and programme contracts.

The right to information and the awareness and knowledge of workers and citizens are fundamental to bringing about the ecological reconversion of industrial activity and the economy, so as to achieve the general objective of improving and protecting the health and safety of workers and citizens. Within this framework it is possible to control, in positive terms, existing conflicts of interest and to strengthen new alliances.

If conditions are to be created to affirm the right of workers to know and to control environmental and industrial risks, and the right of information and of discussion on technological knowledge, all the above should be accompanied by a guarantee of new opportunities for codetermination and acceptance of responsibility by all social parties involved in selecting and defining environmental and ecological contraints.

Environmental cooperation and negotiation facilitates communication between the social partners, fosters the integration of the scientific and political aspects of environmental issues, orientates the consultation process towards the pursuit of solution and promotes joint decision-making. The industrial relations model extended to the environment may promote environmental democracy and industrial democracy.

Bibliography

E. Bartocci. ALLE ORIGINI DELLA CONTRATTAZIONE ARTICOLATA E.S.I. Roma, 1979

T. Treu. L'EVOLUZIONE DELLE RELAZIONI INDUSTRIALI. Spazio-Impresa. Anno 111 n.12, dicembre 1989. E. Procom - Roma.

AA.VV. L'Impresa ambiente n.1 - 1990. L'IMPREGNO ECOLOGICO DELLE IMPRESE ITALIANE Roma.

AA.VV. LA RICONVERSIONE ECOLOGICA DELL'ECONOMIA. Colloqui di Dobbiaco, 13/15 settembre 1990. Dobbiaco - Alto Adige.

G. Fantoni. INTERVENTO ASSEMBLEA ASAP. Roma, 20 luglio 1989.

G. Porta. RAPPORTO SULLO STATO DELL 'INDUSTRIA CHIMICA - VARI ANNI FEDERCHIMICA. Milano

CONTRATO NAZIONALE INDUSTRIA CHIMICA, Capitoli: Salute, sicurezza, ambiente. Vari anni. Ed. ESI Roma.

FILCEA. CONVEGNO NAZIONALE "INDUSRIA CHIMICA ED-AMBIENTE". 1986 - Roma.

FEDERCHIMICA - FULC. ACCORDO SULL AMBIENTE DEL 3-11-83. in appendice a Contratto Lavoro Federchimica 6-12-86- Milano

CGIL - CISL - CISL Milano. SALUTE E AMBIENTE DI LAVORO. Mazzotta 1976 - Milano

Emilio Gerelli. ASCESA E DECLINO DEL BUSINESS AMBIENTALE Dal disinquinamento alle tecnolgie pulite Societa Editrice Il Mulino

Luisa Betri e Ada Gigli Marchetti. SALUTE E CLASSI LAVORATRICI IN ITALIA DALL 'INITA' AL FASCIMO. Franco Angeli Editore 1982.

Scansetti Giovanni. INTRODUZIONE ALL 'IGIENE INDUSTRIALE 1980. Cortina (Torino) Chiuro G.A. PRECANCEROGENESI E TUMORE PROFESSIONALE. 3 Voll., viii, L66.000

ATTI DEL CONGRESSO INTERNAZIONALE SULL'AFFIDABILITA' DEGLI IMPIANTI CHIMICI. Fast milano (P.le Morandi, 2 - 20121 Milano)

Filippo Salvia. IL MINISTERO DELL 'AMBIENTE. La nuova italia Scientifica.

Celli G. LA MINACCIA DEI PESTICIDI. Franco Muccio Ed.

Governa M. LA NUOVA LEGISLAZIONE AMBIENTALE. Maggioli, Rimini.

Maglio S. Santoloci M.   IL CODICE DELL'AMBIENTE. Ed. La Tribuna, piacenza 1989.

PER PENSARE ALL AMBIENTE. Arcadia, Milano 1988.

Cannata G. I FIUMI DELLA TERRA E DEL TEMPO ANGELI, Milano 1987

Sorzoli G.B. LA FORMICA E LA CICALA. Editori Riunti, Roma 1982

Sorzoli G.B. IL PIANETA IN BILICO. Garzanti 1989

MINISTERO DELL AMBIENTE. Relazione sullo stato dell ambiente 1989 Roma

G. Ruffolo. LA QUALITA SOCIALE. Laterza 1985 - Bari

UIL, Anno Teutsch e altri.   LE PORTE DEL SENTIRE. CONVERSAZIONE ECLOGICA DEL SINDACTO. Supplemento a : UIL-SGK INFORM. Bolzano

AA.VV. CONFINDUSTRIA - Roma, 27 - 28 giugno 1990. AMBIENTE: NORMATIVE EUROPEE E PROBLEMI COMUNI

LEGA AMBIENTE. Documento congressuala: il punto di svolta Novembre 89, 3 Congresso, Siena.

AA.VV. Convegno CEEP-AMBIENTE. Roma 1990

INDUSTRIAL RELATIONS AND THE ENVIRONMENT

NETHERLANDS

by

Drs. C. G. Le Blansch

TABLE OF CONTENTS

Page

1. INTRODUCTION.................................... 77

2. THE LEGAL FRAMEWORK............................ 78
2.1 Introduction
2.2 Environmental legislation
2.3 Company environmental protection systems........ 80
2.4 Legislation on the working environment.......... 83
2.5 Legal means for co-determination on
    environmental issues 84
2.6 Discussion...................................... 86

3. VOLUNTARY AGREEMENTS BETWEEN THE INDUSTRIAL ACTORS
   88
3.1 Introduction
3.2 Agreements at national level
3.3 Collective agreements at branch and company
    level... 90
3.4 Involvement and co-determination at company
    level... 93
3.5 Discussion...................................... 95

4. PROGRAMMATICAL STATEMENTS, DEMANDS AND CAMPAIGNS
   97
4.1 Introduction
4.2 Management and Employers' organisations
4.3 Workers' organisations and representatives..... 99
4.4 Programmatical statements at company level...... 103
    4.4.1 The advocates
    4.4.2 Arguments from the workers'
          perspective............................ 104
    4.4.3 Arguments from the environmental
          perspective............................ 106
    4.4.4 Arguments from the functional perspective
          107
4.5 Discussion...................................... 109

5. SUMMARY AND RECOMMENDATIONS 112

   REFERENCES...................................... 115

1. INTRODUCTION

This report gives an overview of the different ways in which industrial relations and the environment are interconnected in the Netherlands. In other words, this study deals with the measures promoted and/or taken within the realm of industrial relations aimed at protecting the environment, modifying the protection of the environment or modifying industrial relations in order to take account of the environment. In this context "industrial relations" are considered to be "the relations between the employers, the workers and their respective organisations and between these groups and the government, as far as these relations exercise a structuring influence on the position of the factor "labour" in industry and society" (Leisink, 1989, p. 5).

In order to provide the most comprehensive treatment of the subject possible, the report addresses itself to the following aspects: (i) Dutch legal conditions in which environment and industrial relations affect each other; (ii) voluntary agreements between employers' and workers' organisations, at national, branch and company level, to the extent that these are concerned with environmental issues and (iii) the policies, programmatical statements and demands from industrial actors. Each chapter ends with an assessment of the subject under discussion. The report concludes with a brief summary and some general recommendations.

2. THE LEGAL FRAMEWORK

2.1 Introduction

Dutch legal conditions under which industrial relations affect or may affect environmental issues and vice versa can be subdivided into (i) environmental legislation, (ii) government-sponsored self-regulation at company level (with the use of financial incentives as well as the threat of legal sanctions) and (iii) legislation on working conditions. These three elements together form the framework for both employers', and workers' and their representatives' involvement in environmental issues. Until now environmental legislation has not extended to any significant degree into the realm of industrial relations. Traditionally workers' and union rights have not been greatly concerned with environmental issues. But this may change with the government's increasing interest in the potential role of the social partners in this sphere.

2.2 Environmental legislation

The Dutch government's environmental policy has acquired a much greater urgency in recent years, as the seriousness of environmental problems has become more visible and tangible. A former minister for the environment has divided the development of Dutch environmental policy into five phases (Winsemius, 1986, p. 34). The first phase was the pre-industrialisation period (until 1900). At this time hygiene and food production were the main concerns and the only environmental legislation was a Nuisance Act (in the Netherlands this Act dates from 1875). The second phase (from 1900 until 1960) consisted of a strong "concentration" of society, particularly in the Netherlands, the country with the highest population density in Europe, the most intensive exploitation of natural resources in agriculture and located downstream from some of Europe's most important and most polluted rivers. At this time policy was mostly concerned with public health and nature conservation.

In the third phase (from the 1960s until the 1980s) there was an attempt to ensure economic growth while at the same time limiting its impact on nature and the environment. There was growing awareness of the finiteness of natural resources, expressed through publications like "Limits to Growth" by the Club of Rome and "Silent Spring" by Rachel Carson. At this time the concept of "public health" was broadened to "environmental hygiene" and physical planning became predominant. The environmental laws emerging in this phase (for an overview, see Tweede Kamer, 1988-89, 21137, 1-2, p. 40) try to deal with polluting activities from different angles, such as:

- the environmental medium concerned (e.g. the Act on soil protection; the Act on fresh water contamination; the Act on sea water contamination);

- the type of pollution (e.g. the Act on noise pollution; the Act on waste disposal; the nuclear energy law; the Act on chemical waste disposal) or

- the source of contamination (e.g. the Nuisance Act, the mine law, the Act on the prevention of pollution by ships).

Because of this a situation arose in which laws partially overlap and environmental legislation is sometimes confusing, being sometimes contradictory and enforced by several different authorities demanding widely differing action. This in turn required more coordination, which was provided by the enactment of a general Act on environmental hygiene (Wet Algemene Bepalingen Milieuhygiëne, 1980). This Act is eventually intended to become an overall general environmental law, thereby fitting in with Winsemius' fourth phase.

This fourth phase, which began around 1980, Winsemius calls the "integration" phase, in which "environment" is the key concept and a goal in itself and is still continuing today. The fifth phase, that of "survival", is posited for the future. At global level this phase is epitomised by the Brundtland Report, "Our Common Future". In the Netherlands a similar role is played by the central governmnent institute for public health and environmental hygiene (RIVM) report "Zorgen voor Morgen" (Caring for Tomorrow) and the government's subsequent key publication on environmental policy, the National Environmental Policy Plan (NEPP) (Tweede Kamer, 1988-89, 21137, 1-2), later updated by the so-called NEPP Plus (Tweede Kamer, 1989-89, 21137, 20-21). The NEPP outlines the strategy for environmental policy in the medium term. The report states that the strategy has been developed against the background of the desire to solve or bring under control environmental problems within one generation. It sets out the policy actions which are deemed necessary during the period from 1990 until 1994 in order to begin in the desired direction.

The report describes a number of instruments at the disposal of the government which could be used to influence the behaviour of business and industry and the public: voluntary and other agreements, legal regulations, financial incentives, information, setting a good example and penalties (Tweede Kamer, 1988-89, 21137, 1-2, p. 12). It also describes procedural safeguards for the drawing up of voluntary agreements.

The government wants to play an active and encouraging role and in doing so wants to employ the most effective means available, having reached agreement on them as much as possible through consultation with target groups such as groups of producers and consumers, government agencies and other concerned parties. Where regulation is necessary, the government believes it should be the end result of a consultation process.

The year the NEPP was published, 1988, can be regarded as the point at which government began treating the environment as an issue concerning industrial relations and vice versa. From then on, unions and employers' organisations have been seen as part of the aforementioned target groups, to be consulted and encouraged to become involved in environmental policy making. In the NEPP the government calls on the unions to help implement an environmental policy aimed at securing sustainable development (given their position and their responsibilities as contributors to environmental aspects) in consultation with employers.

A principal aspect being considered is the setting up of "internal environmental concern systems" (also called "corporate environmental protection systems" which are examined below) and encouraging members to take tangible steps to implement environmental policy. It is stressed that the role of the unions is important not only at company or branch level but also internationally. Employers' organisations are urged to work for sustainable development, both within their organisations and through their contacts nationally and internationally (Tweede Kamer, 1988-89, 21137, 1-2, p. 229).

In the following year, 1989, the government issued a memorandum entitled "Company Environmental Protection" (Tweede Kamer, 1988-89, 20633, 2-3). Since this report deals with environmental management and hence in-company affairs, it devotes considerable attention to the role of workers and workers' organisations. Since employers' organisations had in fact already played a leading role in the formulation of the central concept of the report, its origins and contents merit a closer look.

2.3 Company environmental protection systems.

In the course of the 1980's the Dutch government realised that regulation alone does not produce an effective environmental policy. It realised that to counter further deterioration of the environment, it would have to rely on the cooperation and preventative actions of other social actors. The so-called "internalisation drive", intended to persuade producers and consumers to internalise environmental standards and values, has been an integral part of government environmental policy since the publication of the "Indicative Multi-Year Programme on the Conservation of the Environment 1985-1989". Here the government described its task not so much as one of regulating as one of encouraging.

Employers' organisations were quick to react to this development. In a publication jointly produced by the

Federation of Dutch Industry (VNO) and the Netherlands Christian Federation of Employers (NCW), "Environmental Protection in Companies" (VNO/NCW, 1986), the two organisations proposed the creation of an environmental protection system in companies. The purpose of such a system would be "to ensure better compliance with environmental laws and regulations". The system is clearly presented as an alternative to "more government intervention and yet more regulation in this sphere" (VNO/NCW, 1986, p. 5).

In the Dutch language the term "company environmental protection system" ('bedrijfsinterne milieuzorgsystemen') tends to cause some confusion, since it suggests that it is referring to protection of the environment within the company. But the reason for this nomenclature must be sought in the course of its development. Environmental protection was initially a government matter and a company's pollution was inspected by the government. Only later was a system suggested in which environmental protection would also become a matter for companies, and companies would themselves control and manage their own pollution. It should be noted, incidentally, that the internal and external company environment and its protection, can of course be closely linked.

Following on from the system proposed by the employers' organisations, a commission was set up to investigate the desired and possible relationship between the government and companies with environmental protection systems in place, and to develop criteria for measuring how well such systems were functioning (Commissie Bedrijfsinterne Milieuzorgsystemen, 1988). The Commission on Company Environmental Protection Systems, a temporary research committee composed of government and employer representatives which advises the Minister for the Environment, formulated what it called a "standard company environmental protection system".

This system, which enjoys undisputed status in the Netherlands, consists of eight elements:

- a written environmental policy statement;
- the integration of environmental protection into the operation of the company;
- an environmental programme;
- a measurement and recording programme;
- an internal control system;
- internal and external reporting;
- internal information and training; and
- a regular audit of the company's environmental protection system.

The commission recommended that the government encourage the development and implementation of these systems and help develop model systems for particular groups of comparable companies.

In "Company Environmental Protection" Ed Nijpels, then Minister for the Environment, outlined his plans for the promotion and internalisation of responsibility for environmental matters in companies. In this paper the minister accepted in full the ideas previously articulated by the employers' organisations. The plan was for all large companies causing high or medium levels of pollution or with special environmental risks (between 10,000 and 12,000 companies) to have environmental protection systems in place by 1995. Smaller companies subject to the Nuisance Act (around 25,000 in total) were also to have taken steps towards a partial environmental protection system by then.

To achieve these goals the government opted for incentive subsidies. But the introduction of Company Environmental Protection Systems is only seemingly voluntary since prescriptive government rules are already being prepared. If it emerged from the planned interim evaluation of progress at the end of 1992 that voluntary means were proving inadequate, then the government would take recourse to legislative means.

In the meantime a programme of measures was presented with the aim of encouraging the development of company environmental protection systems (as recommended by the commission), disseminating information on this issue to target groups, companies, workers and government bodies, and financing professional and vocational training and specific research. More detailed programmes were subsequently presented (Ministerie VROM, 1990; Ministerie VROM, 1991).

These programmes illustrate the thrust of government policy. It devotes resources to environmental protection projects, to information programmes organised by trade unions and employers' organisations, and to the creation of "corporate environmental services" (bedrijfsmilieudiensten). These are institutes, as yet still being planned, whose purpose will be to support in particular smaller companies in developing their environmental policies. It seems that corporate environmental services are to be organised along similar lines to the corporate health services provided for by the Working Environment Act (see e.g. van den Berg and de Raaff, 1990).

As far as workers' interests are concerned, the memorandum "Company Environmental Protection" stresses in particular the relationship between the environment and working conditions, both with regard to the way in which matters are organised within the company and to their content. Where these relations are close, it is suggested, legislation on working conditions should have an effect on the environment as well.

2.4 Legislation on the working environment

The most important legal instrument in the Netherlands on the quality of working life is the Working Environment Act (WEA, known in Dutch as the "Arbowet"). Intended to increase the level of safety in the workplace and to improve both the physical and mental health, and wellbeing of workers, this Act was adopted by Parliament in 1980, but was not fully enacted until ten years later. Since 1990 it has covered all persons working for an employer, both in the private and public sector, in large corporations and in small firms. The WEA is primarily an enabling Act - essentially a framework which provides the basis for supplementary decrees. (Bayens and Prins (1990, p. 15) give the example of the minister having the power to issue decrees on, say, asbestos).

A second important aspect of the WEA is that it defines the roles of the minister, labour inspectors, employers, workers, works councils, safety and health specialists and the Working Environment Council (a committee of the works council with specific legal rights to deal with matters of safety, health and well being). Each are given responsibilities, rights and duties concerning the working environment. The WEA also regulates the structure of the occupational health and safety system within companies and the organisation of public supervision and monitoring.

The WEA sets basic standards on both the physical and psychological aspects of the working environment. The aim of the Act is not only to ensure suitable standards of health, safety and wellbeing (the latter being seen as an effect of "the quality of working life" and related to job structuring, elimination of repetitive tasks, inter-colleague contacts and the prevention of alienation). It also places duties on the employer to organise work in such a way that the health, safety and wellbeing of the workers are assured to the highest possible level. Although the employer is responsible for the working environment, workers have certain responsibilities for their own safety and health and are obliged to cooperate with the management on health and safety matters.

Some of the WEA's provisions relevant in this context are the following:

- Employers with more than 100 employees are required to prepare an annual written plan of their safety, health and wellbeing policy;
- The policy on health, safety and wellbeing as well as the annual plan must be presented to and discussed with the works council;
- Certain factories, particularly in the chemical industry, are required to prepare safety assessment reports;

- Industrial enterprises with more than 500 employees are required to have a safety service attached to them. (According to Bayens and Prins (1990, p. 43), smaller firms often have a part-time safety technician, while firms with fewer than 100 employees as a rule have no safety specialist at all.) The safety service or the safety expert makes recommendations to both the employer and the works council.

The WEA provides the means to issue decrees applying to company, branch or national level but does not itself contain detailed requirements. The WEA, "taking notice of the best available technical practices, the state of the art of company health care and scientific knowledge in the sphere of ergonomics as well as labour and management science", aims to obtain "maximum possible safety and maximum possible protection of health, as well as the furtherance of and maximum possible care for the wellbeing of the worker, unless such cannot be required in reason" (WEA, article 3, sub 1a).

Given this field of action the question arises as to whether and to what extent the WEA can also offer a framework for dealing with environmental issues. It seems that the answer to this question depends on the nature of the environmental issue at stake and is ultimately left open to discussion and interpretation. Special attention should be paid here to article 11 of the WEA which deals with the prevention of danger to a third party: "When by, or as a result of, labour conducted by workers on instruction of the employer, in a company or an institution or in its immediate surroundings any dangers can arise for the safety or the health of people other than the workers involved, the employer is to take effective measures to prevent such dangers". This article clearly offers a point of application for introducing environmental issues that are not directly concerned with workers' health, safety or wellbeing.

2.5 Legal means for co-determination on environmental issues

Since the applicability of legislation in the field of industrial relations to environmental issues is open to question, it is interesting to note that most of the parties concerned at national level consider the legal means available for co-determination on environmental issues to be sufficient in principle.

In this context reference is made to article 2 of the Works Councils Act, which states that the works council exists in the interest of the company in all its objectives.[1] This

---

[1] Recent research (Mulder and Benders, 1991) has shown that this also applies for co-determination organsin non-private organisations if to a slightly lesser degree.

means that the works council has a legitimate right on paper at least to deal with environmental objectives.

According to the government the Act offers sufficient scope for "active thinking along" (Tweede Kamer, 1988-89, 20633, 2-3, p. 34). Ket (1989, p. 11) also argues that environmental objectives can be achieved through the Works Councils Act and the Working Environment Act. So too does Klatten (1989), in an address to a trade-union conference on works councils and the environment.

The Federation of Dutch Trade Unions (FNV) (FNV 1990, p. 1) has declared itself in full agreement with the government that the two acts offer "sufficient points of departure". The Manufacturing and Food Industry Union of the Christian National Federation of Trade Unions (CNV) (Industrie- en Voedingsbond CNV, 1989, p. 39f.) also agrees, and has provided an overview of the possibilities offered by the existing legislation.

Most of the above protagonists will admit that at times a fairly broad interpretation of the legal texts is necessary. For instance, in article 25, sub 1h of the Works Councils Act, which deals with the mandatory consultation request on investments, the council has to take it upon itself to assess investments on environmental criteria. And while the Working Environment Act offers several options for co-determination in explicitly mentioned spheres such as safety, health and wellbeing, in the sphere of environmental protection this remains at least at present implicit.

Currently two points are still under debate:

(1) Regarding the description of aspects which the works council is entitled to supervise, the FNV calls for the inclusion of "environment" in article 28 of the Works Councils Act. "The addition of responsibility in the sphere of the environment will give the works council an additional legitimation to put environmental issues on the agenda at any time" (FNV, 1990 p. 7).

(2) The workers' organisations call for the institution of the employee's explicit duty to report environmental transgressions as well as the right to suspend work in case of acute environmental danger. The government's "Company Environmental Protection" says the following on this issue: "With regard to reporting as such and the accompanying legal protection, the government proposes to request the advice of the Social Economic Council (SER) before formulating a view. The same applies to the right to suspension of work at times of environmental risks with immediate danger for the vicinity of the company".

This request for "advice on the involvement of workers in company environmental protection" was put to the SER, the tripartite body which advises the government on economic issues, on 5 April 1990.

2.6. Discussion

The above observations give the impression of a field on the move, in which on the one hand public awareness of and government interest and intervention in what is happening in companies are growing, while on the other hand the workers' right to co-determine company policy is steadily being extended into the environmental sphere. However, some qualifying remarks are necessary.

Although public interest and intervention in what is happening in companies are definitey growing, the consequences for workers' involvement are not yet certain. Public attention breeds formalisation and centralisation (compare Mintzberg, 1979, 288f.), as is the case in the emergence of the Company Environmental Protection System, which in fact is a formalised management system. Because of this formalised and centralised nature, workers' involvement with the system is not self-evident.

And the extension of the workers' rights to co-determine company policy into the environmental sphere is one thing; a co-determination practice on environmental issues is another. As far as the instruments for co-determination are concerned, some inadequacies have been shown in practice. Although almost everyone in the field states that the legal instruments are in place, there are still deficiencies that should be pointed out. Firstly, the legal instruments are only available to companies where a works council is in existence. This means that a number of options are not available to workers employed in smaller companies with fewer than 100 staff, and that none are available to those in companies with fewer than 35 staff. It should also be noted that works councils have not been set up in a considerable number of companies which are legally required to have them.

Secondly, the usefulness and effectiveness of the legal instruments is restricted. For instance, the powers at the disposal of a works council to enforce the various provisions of the law vary enormously. Means of appeal are often difficult to pursue, especially in the light of fact that rights concerning environmental issues are mostly left implicit. Severe problems can arise here for works councils in conflict with the employer.

For example, the right to information (guaranteed under article 31 of the Works Councils Act) can easily become a hollow shell where environmental issues are concerned, since the Supreme Court has ruled that works councils cannot simply demand information by invoking their general right to monitor developments within their company, and that they will have to refer to explicitly granted rights under the Act or other regulations (SoZaWe, 1987, pp. 153f.). It seems likely, therefore, that the use of, or readiness to resort to, non-juridical pressure determines the effectivenessof workers' input at least as much (see Teulings, 1981).

In short, the right to co-determination with regard to the company's environmental policy has its restrictions. But an overly detailed review of the legal provisions and imperfections of environmental co-determination carries the danger of ignoring the fact it is not only the comprehensiveness of legislation which determines workers' involvement and co-determination.

Consciousness and readiness of both workers and their representatives and employers are required in order to set off a dialogue, in which many problems and obstacles may still arise. Earlier experiences with attempts to realise involvement and co-determination on the introduction of new technologies indicate it might well be prudent to temper somewhat the exaggerated expectations about workers' influence on environmental policy, certainly in the short term.

The same goes for the still limited degree to which the Working Environment Act is applied today and the recent somber assessment by Bert de Vries, the minister for social affairs and employment.[2] Since at this stage one can hardly claim that such established forms of the right to co-determination are institutionalised, the introduction of new forms might require a veritable tour the force.

The next chapter will deal with voluntary agreements on the environment between the industrial actors at national, branch and company level. Also attention will be given to the actual measure of involvement and co-determination at company level concerning environmental affairs.

---

[2] The legal measures indtroduced so far are insufficient, as has been shown in practice. Works councils do not get involved in working conditions, the workers know nothing about it and even middle management is hardly involved in changes (Bert de Vries, quotedin De Vilkskrant, 4 May 1990).

3. VOLUNTARY AGREEMENTS BETWEEN THE INDUSTRIAL ACTORS

3.1 Introduction

Social partners can agree more or less extensive procedural or substantial arrangements to reduce environmental pollution and increase the environmental awareness of their members and affiliated organisations. This chapter provides an overview of voluntary agreements in the sphere of industrial relations concerning the environment, at national, branch and company level.

3.2 Agreements at national level

At national level one form of consultation between employers' and workers' organisations has become institutionalised in the aforementioned tripartite Social Economic Council (SER). This body advises the government on economic issues, and in recent times has also become increasingly involved with environmental questions. Advice has been given, for example, on the Brundtland report, the National Environmental Policy Plans (NEPP and NEPP+) and on topics such as "an environmental hallmark" and "domestic waste disposal". The SER-committee on company research (COB) published a manual on commuter traffic (COB-SER, 1991).

Recently the SER published a report by its "committee of economic experts" (CED) on Environment and Economy (SER, 1991), in which an extended use of market-conform instruments like regulating levies and marketable emitting rights is recommended.[3] Here reference is made to a system under which a sector of industry obtains the right to cause a certain amount of pollution. Individual companies receive a part of this right which may be traded between companies. A company which reduces its emissions can therefore sell part of its right to other companies that want to pollute more than their right. In this way the environment has a price and pollution prevention pays.

But also specific negotiations on the environment have taken place. In January 1989 the FNV, the largest Dutch trade union and the VNO, the largest employers' organisation, reached an agreement on a common approach to environmental problems. In their joint statement the FNV and VNO declared that an environmental policy is necessary even at times of slow economic growth. This declaration, made against the background of growing concern for, and awareness of, environmental problems started a process of

---

[3] This refers to a system which a sector of industry obtains the right to cause a certain amount of pollution. Individual companies receive a part of this right which mat be traded between companies. A company which reduces its emissions can therefore sell part of its right to other companies that whanto to pollute. In this way the environment has a price and pollution prevention pays.

negotiations on the environment between all employers' and workers' organisations (including the Christian employers' and workers' organisations (NCW and CNV) and others).

The starting points of these negotiations were that economic growth should be sustainable, that government should play a key role in environmental policy, and that potential disadvantages in terms of reduced international competitiveness should not be an excuse for reticence or inaction. Another starting point, also mentioned in the declaration, was the recognition of the role of trade unions in environmental matters at all levels, including the company level.

As a result of the agreement different working groups have been established. One such group is studying the government's NEPP, another is investigating the problems of commuter traffic, a third is formulating recommendations for the improvement of public transport to an from work, and so on. Another result of the cooperation was the initiative to launch a campaign on "environmental protection within companies". This initiative was based on a memorandum on environmental protection which dealt with the opportunities for joint approaches (Milieuoverleg RCO / Vakcentrales, 1989). According to the memorandum, the important thing is to try to ensure as quickly as possible, as systematically as possible and in as many companies as possible that environmental aspects are taken into account in all facets of the production process.

The paper presents an action programme, which distinguishes three levels:

(a) **National level**
Employers' and workers' organisations are able to play an important role in raising the level of awareness of environmental problems within industry. They can also focus the attention of their members and affiliated organisations on aspects that require further agreements between the participants on the development and implementation of company environmental protection systems.

(b) **Branch level**
This level is given a pivotal role both with regard to increasing environmental awareness and in taking the initiative for the development of a branch-specific company environmental protection system. Furthermore, the employers' organisations at branch level are considered to be the appropriate bodies for providing information and instruction on the system that has been developed and for persuading individual companies to adopt it. Workers' organisations should play a part in designing and setting up the system and framing instruction programmes. The practical detail of the system should be left to the parties involved at branch level.

(c) **Company level**
The ultimate aim of the action programme is the introduction of environmental protection systems in as many individual companies as possible. To be successful the system will have to be supported wholeheartedly throughout the company. Government regulation in this context is rejected. Moreover, the successful introduction of the system requires the involvement of the whole staff. Consultation with the works council on the procedures and opportunities for introduction is desirable, as is the use of existing communication channels within the company.

Two specific activities were announced to initiate the actions outlined above. One was the launch of the national campaign on environmental protection, aimed at affiliated organisations and provincial and local authorities. The idea was that after this event, organisations at branch level would take over the torch and start to unfold activities within their own domain. The second activity consisted of further deliberation with government on the formulation of an appropriate programme to promote the introduction of environmental protection.

Continuing the programme outlined above, the environmental protection campaign took place in September 1989. The joint consultations on environmental issues at national level took place as announced, but did not exactly bloom. A communique issued in January 1990 referred to criticism from the workers' organisations regarding the functioning of the environmental consultation (milieu-overleg).

This makes mention of an exchange of views, after which it was agreed to continue the consultations, on the basis of two new objectives:

(i) to discuss each other's views on environmental issues in order to reach a better mutual understanding; and

(ii) to investigate specific environmental matters on which joint activities could be undertaken.

Since the issuing of the communique no official public statements have been made. Informal comments are, however, to the effect that the consultation process is not functioning as intended. The question is whether these consultation practices will last without some tangible results in the short term. It is understood that an evaluation will take place in the near future.

3.3 Collective agreements at branch and company level

At branch level trade unions are involved in two kinds of consultation on environmental issues. Firstly, within the framework of consultation between the government and so-called target groups. The main parties involved in this consultation are the environmental authorities on the one hand and representatives of industrial branches on the other. In the construction industry and manufacturing industry unions also take part in these discussions, the aim of which is to establish "covenants". These are special agreements between the government and groups of companies in which unions are not supposed to be a contracting party.

Secondly, there is the process of collective bargaining, both at branch and at company level, in which the environment has become a topic of discussion, negotiation and agreement in recent years. In fact, collective agreements are among the principal areas in which union policies on the environment are realised.

The smaller Christian trade unions in particular stress the use of collective bargaining in dealing with environmental issues in the context of industrial relations (Industrie- en Voedingsbond CNV, 1989). In a memorandum on the state of its environment policy as of May 1990, the CNV Manufacturing and Food Industry Union drew up a balance sheet on the introduction of the environment in recent collective bargaining (Industrie- en Voedingsbond CNV, 1990). The main aim of the union's collective bargaining policy on the environment is to reach agreements on the participation of workers in corporate environmental decision making.

The memorandum reported that 45 out of a sample of 65 collective agreements in the manufacturing and food industries in place in May 1990 addressed the question of the environment in some form or another. In almost all cases this was the first time that an environmental clause had been included. The clauses differed widely, however.

The report classifies the agreements as follows:
(a) Four agreements made mention of "the development of environmental policy in cooperation with works councils and trade unions".
(b) Twenty-four agreements contained arrangements acknowledging the role of the union, in the form of obtaining information or discussing environmental issues in the so-called "periodic consultation" (between unions and management).
(c) Six agreements made mention of management's intention to inform the works council.
(d) Branch agreements in the engineering industry as well as three agreements at company level provided for the setting up of company environmental protection systems.
(e) The remainder was a diversity of agreements concerning, among other things, training programmes, the appointment of environmental experts or "strengthening workers' rights".

In 1991 the Collective Bargaining Service, an official body which registers the results of collective bargaining and assists in the task of making agreements binding for all parties concerned, published a report on agreements concluded up to April 1991 which included an analysis of clauses concerning the environment (Dienst Collectieve Arbeidsvoorwaarden, 1991).

It defined "environmental clauses" as "agreements concerning the external environment", but did not take into account agreements presented exclusively in the context of working conditions (p. 21). Environmental clauses were included in 53 of the 161 collective agreements studied.[4] In most cases the environmental clauses were agreed after 1989. Before that year only seven agreements contained such clauses. The report divides the environmental clauses into four categories:

(a) Agreements providing for consultation with workers' organisations and/or research and development on environmental issues (22 in total). This group also contains the agreements outlining statements of intent. As an illustration, the report mentions the collective agreement for the engineering industry, which contained an undertaking that job structuring would be carried out in such as way as to minimise the burden on the environment. In agreements for cleaning companies the employers undertook to explore substitution with less-polluting materials.

(b) More wide-ranging agreements (12 in total). Agreements in this category went one step further providing, for instance, for the introduction of company environmental protection systems, or a system of environmental reporting or discussion of environmental data.

(c) Agreements containing concrete measures (14 in total). This category includes two detailed agreements from the Philips company, in connection with agreements it made with the Ministry of the Environment in 1986. Several agreements in the construction industry provided for the banning or using up of a range of hazardous materials, such as asbestos. The agreement for the building trade contained an updated clause providing for the recycling of tar. Clauses on the banning of asbestos are also found in agreements covering car and tyre manufacturers. The Shell Oil company agreed to improve the means for individuals to raise environmental issues and to consult workers' representatives, works councils and/or trade unions on environmental policy and annual plans where appropriate, and if necessary to take action. Information, training and the availability of experts were also mentioned.

Haulage firms undertook to set up a bipartite committee to make recommendations on the issue of drivers' liability for toxic waste and work which damage the environment. The agreement for Paktank referred to workers' obligations to cooperate fully with the employers' efforts to reduce

---

[4] Here it should be noted that this survey extended beyond the manufacturing and food industries

pollution to a minimum and to follow rules and procedures to the letter. And the biochemical company Gist-Brocades agreed not to bring into commercial exploitation any new processes whose safety aspect and environmental impact had not been sufficiently studied and controlled to reduce the risks to the workers involved and reduce their damaging effect on the environment to a minimum.

(d) Agreements dealing with commuter traffic (6 in total). These agreements provided for an employer's obligation to employ workers within a 30-kilometer radius of their home (in the case of private security firms), to pay the transport costs for workers that need to travel frequently (cement and cement-transport companies), and to take into consideration the union's proposals for reducing environmental pollution as a result of commuter traffic (the chemical company AKZO).

3.4 Involvement and co-determination at company level

As discussed earlier, in recent years there has been a growing emphasis on company self-regulation and the introduction of company environmental protection systems. This has in turn caused more attention to be focussed on what is actually happening inside companies concerning workers' involvement in environmental protection. Until recently only general impressions and incidental practices in this area have been available.

Some of these can be found in a report by the Commission on Company Environmental Protection Systems (1988). It made an inventory of the consultative structures in which environmental issues are discussed and found that in half the companies studied (9 in total) consultation took place at the lowest level, "on the shopfloor". The commission also noted that "in several companies consultation occurs in special committees, the works council and other bodies, which must remain outside the scope of this report" (Commissie Bedrijfsinterne Milieuzorgsystemen, 1988, p. 24f.).

The government has also outlined its general impression on this in "Company Environmental Protection": "The government notes with approval that at a practical level a tendency can be observed that, depending on their nature, agreements are concluded at the level of the works council, within the collective bargaining process and in the regular consultation between employers and employees" (Tweede Kamer, 1988-89, 20633, 2-3, p. 34).

Much of the information on case studies and other publications in this field originate from educational institutions, which are not intended to paint a reliable picture of current practice. A study of the Environmental Education Foundation, a private commercial organisation, (Hengelaar, 1989) concludes that "most works councils have no doubt that monitoring the company's environmental policy is part of their brief" (Teijlingen, 1989).

But since only 22 of the 98 works councils approached in the context of this study replied, it is difficult to draw more general conclusions from these findings. Of the 22 works councils that responded, 12 pursued an environmental policy. They did so by means of raising environmentally damaging situations with the management (all 12), supervising the company's environmental policy (10), assessing investments (8), proposing improvements in management (5) and preparing an environmental memorandum (1).

Only recently has more reliable material on this issue become available. The first in line was a representative study by the Manufacturing Union FNV (1989), in which a considerable number of workers employed in the chemical industry in the Rotterdam area were asked about environmental and other issues. Asked which measures should be taken in the environmental sphere, 40% said that the dissemination of information to staff should be improved. A majority of those questioned were of the view that workers and unions had an important role to play concerning the environment. The report notes that it was remarkable that in open questions on the environment 12.4% of the respondents suggested ideas for improving its quality.

In 1990 the same FNV Manufacturing Union conducted research on the activities of works council sub-committees on working conditions (provided for under the Working Environment Act). The activities of these committees were ranked according to prevalence. Out of 22 issues identified, concrete environmental issues ranked relatively high:
- Second: noise (82% of committees had dealt with noise problems);
- Third: hazardous materials (79%);
- Twelfth: effluent, waste disposal and emissions (47%; and
- Thirteenth: leaks and spillages (47%).

Four out of the ten committees claimed they had been successful and achieved results on these issues. However, one of the people conducting the research noted that "in practice many environmental issues are intertwined with health and safety issues. A leaking container of toxic chemicals is harmful to both the environment and the health of workers. Moreover, often conflicting interests are involved. For workers it is better to have a noisy compressor situated outside on the roof; but that is unpleasant for people in the vicinity" (Praktijkblad Medezeggenschap, 1990, p. 22f.).

Particularly interesting is the research commissioned by the Ministry of the Environment to evaluate the introduction of company environmental protection systems

(Calkoen and ten Have, 1991). In addition to a number of environmental protection characteristics, this report deals also with the declared involvement of staff in environmental protection. Some 46% of companies claimed that their workers had specific environmental tasks, and 18% claimed that staff were involved in setting up an environmental programme. No less than 40% of the companies interviewed claimed that environmental issues were an important element in discussions with the works council, and 46% said that these issues were discussed at shopfloor meetings.

A problem that arises here is that data on the involvement of workers and works councils are based on interviews with managers and environmental coordinators and are therefore of limited validity. An even more serious problem is that it is not clear what involvement of works councils actually means in practice. Does it reflect an active environmental discussion throughout the company? Or do works councils take initiatives when managements fail to do so?

Some light is shed on these questions in a forthcoming report on an exploratory research project on workers' involvement in environmental issues (Le Blansch, 1991b). This distinguishes between "functional" co-determination on the one hand and "strategic" co-determination on the other.

Functional co-determination on environmental issues, which remains within the boundaries of the chosen company policy, can occur under conditions of adequate rights, adequate industrial relations structures and access to information. Strategic co-determination, which questions the company's environmental policy, occurs under conditions of sufficient power and environmental awareness among the workforce. The report concludes that in the current Dutch legislative context, strategic co-determination on environmental issues asks a lot of the workers, perhaps even too much.

3.5 Discussion

The above shows that the interest of employers and unions in the environment has led, especially since 1989, to a growing number of agreements on, and stronger involvement with environmental issues at both national, branch and company level. Some qualifying remarks may put this finding in perspective.

In principle there need not be any conflict between employers and workers concerning environmental protection. In fact, joint approaches help matters considerably. Nevertheless, there is only a limited willingness on the part of the employers to make common cause on this issue. This reluctance appears to arise from the general tendency with social partners to analyse the behaviour of the other side (i.c. the unions) in terms of power only.[5]

In relation to this it is interesting to note that Windmuller et al. make mention of a traditional hostility of Dutch employers to snoopers of all kinds, in particular trade unions (see Windmuller, de Galan and van Zweeden, 1983). This may also be the reason, despite all the ringing declarations, behind the deadlock in the consultations at national level. Another reason for this deadlock may be the difference in style between the employers' environmental bureau (used to lobbying) and the trade unions at the national level (used to negotiating).

In addition, some remarks can be made on the contents of the agreements. It is noticeable that collective agreements containing concrete environmental provisions often deal with environmental issues in relation to working conditions. A good illustration of this is the frequent occurrence of agreements on asbestos, even though the government has also taken measures in this area (see the aforementioned example in 2.4). This also seems to indicate that, perhaps to begin with, particularly those clauses which are already widely accepted are included in a collective agreement.

Recent Dutch research suggests that the environment and working conditions are very often treated as interlinked, and not just at the level of collective agreements. Moreover, it is also suggested that procedural arrangements only appear to be "effective" if they are firmly based in a company's everyday practice (Le Blansch, 1991b). These findings suggest that collective agreements as such only play a small part in bringing about changes in environmental practices.

In short it can be said that in the Netherlands the number of agreements on the environment and the intensity of workers' involvement in this sphere is steadily growing. The starting points are the working conditions and everyday practice. As the agreements seem to so far present more a codification of practices than a modification, expectations of the impact of collective environmental agreements should not run too high in the immediate future. However, it is far too early for a comprehensive evaluation.

---
[5] One more demonstration of the tendancy to seethings from a power perspective rather than from a fuctional perspective can be found in a VNO memorandum on collective bargaining. Here the unions environmental activities within companies are placeed in a territorial perspective and are therfore rejected. It states 'When in collective bargaining.demands are made within the framework of the FNV's environmental policy, the following should be considered. The FNV's policy is a clear attempt to provide a role for union work within companies and thier organisations have always maintained that no consultations should take place between management and union member groups. Consultation on any subject is to be exclusively with either the works council or the paid union representative. This principle should be upheld especially since the union is trying to extend its activities within companies in other fields as well. In maintaining this position it is neither necessary or desireable to agree on behalf of union members to protecty their position within the company or to hand over information. (VNO, 1989,p42)

4. PROGRAMMATICAL STATEMENTS, DEMANDS AND CAMPAIGNS

4.1 Introduction

This chapter presents the attitudes of both employers' organisations and trade unions on environmental protection measures, analyses them and places them in their historical context. Statements on workers' involvement at company level are discussed separately, as they reveal many of the assumptions behind the role which workers, unions and co-determination practices can play on behalf of environmental and workers' interests. Again, the chapter is concluded with a discussion.

4.2. Management and employers' organisations

As described in section 2.3 of this report, employers' organisations have played an important role in the development of company environmental protection systems. In this respect 1986 can be seen as a watershed. Before that time, the activities of the VNO, NCW and their joint environmental agency, the Bureau for Environment and Physical Planning (BMRO) were somewhat defensive towards environmental protection, mainly stressing the negative consequences for profits and international competitiveness. These arguments were put forward in two specific publications on the issues of environmental policy and planning (VNO/NCW 1982, 1985).

A shift occurred in 1986 when the employers' organisations adopted a more proactive stance, developed the concept of systematic company environmental protection and started to promote this concept to their members. The result was publications such as "Environmental Protection in Companies" (VNO/NCW, 1986), "Company and Environment" (VNO/NCW, 1988) and "Environmental Reporting by Companies" (VNO/NCW, 1990 and 1991). It is for this reason that the government's "Company Environmental Protection" declared that it is "thanks to the Bureau of Environment and Physical Planning (BMRO) of the VNO and NCW that in the Netherlands attention is focused primarily on the systematic approach [italics KLB] of environmental protection as a question of control, on behalf of which instruments like environmental auditing can be developed and applied. Moreover, the cooperation and interaction between government and industry in developing means for protecting the environment can be considered characteristic for the Netherlands" (Tweede Kamer, 1988-89, 20633, 3, p. 18).

It can be observed that BMRO was very successful in convincing the government of industry's definition of the environmental problem and its possible solutions. In a recently published dissertation, Doorewaard (1990) asserts that the role fulfilled by the BMRO in providing assistance to government and industry has been very succesfull. She describes the task of an intermediary organisation like the BMRO as "controlling contextual changes in such a manner that enterprises can adapt themselves (without the need for major changes of policy)". If it is to function properly, that is, to provide optimum assistance to government and industry concerning long-term solutions to environmental problems, it must adopt a change-oriented attitude in which initiatives are, at the very least, focused on the opportunities presented by profound problems. The BMRO has succeeded in this, which according to Doorewaard can be explained by key characteristics such as close relations with members and the government, taking and pursuing initiatives, and knowledge and expertise (Doorewaard, 1990, p. 197).

As public and corporate awareness concerning the environment continues to grow, a number of studies, from various sources, have been published on the state of environmental protection in companies and the attitude of management. Most studies use their own theoretical and operational definitions of environmental protection. The result is that, for instance, one study published in 1990 concludes that 10% of (smaller) companies are planning environmental investments in the near future (KNOV/AMRO, 1990, p. 18), while another claims that 63% of Dutch companies made "environmental investments", including "indirectly contributory investments", in 1990 (Nationale Investeringsbank, 1991, p. 5). The wide difference between the two figures gives an indication of the problems associated with assessing data on this issue.

Equally questionable from a methodological point of view but interestingbecause of its international comparisons is a Touche Ross study, "European Management Attitudes to Environmental Issues" (Touche Ross Europe Services, 1990). Dutch companies are remarkable for being the only ones among those interviewed with a board member responsible for environmental management.

Half of them have written environmental policies and employ environmental managers, often with teams of two or three experts. The study notes with interest that most Dutch companies do not expect any major impact from future legislation and none expect the European Community and 1992 to have any impact. This may be, according to Touche Ross, because they believe that the Dutch level of environmental protection is already ahead of the rest of Europe. This would appear to be supported by the fact that only one of the companies interviewed claimed to having changed its production processes, and two their products, as a result of new legislation. The others declared that they always made changes ahead of legislation being passed. In the long term, most companies planned to improve their environmental performance regardless of legislation, the study concludes.

Two recent publications deal with the degree of penetration of company environmental protection systems in relation to the government intentions in this regard. One is yet another BMRO study, an assessment of company environmental protection systems in several branches of the economy (VNO/NCW, 1991). This contains qualitative information on fairly diverse activities, and therefore does not provide a comprehensive overview.

The other study is the aforementioned representative survey on company environmental protection systems commissioned by the Ministry for the Environment (Calkoen and ten Have, 1991). According to this report, 4% of Dutch companies said they had such a system in operation, while another 26% claimed to be developing one. The highest degree of penetration is in the chemical industry (70%), followed by the food-processing industry (47%). The construction industry has been the least active, with only 20% of companies setting up a system. Company size appears to be a determining factor with larger companies in particular are working on an environmental protection system.

It is very interesting to note that in trying to correlate company characteristics, the authors found a significant link between the number of elements of the system [6] on the one hand (more or less representing the quality and degree of formalisation of environmental protection) and the presence of a works council or a plan for improving working conditions. They note that even when the data are corrected for company size and branch of industry, the linkage remains statistically highly significant.

4.3 Workers' organisations and representatives

A 1990 issue of the journal Tijdschrift voor Arbeid en Bewustzijn (an independant part-University staffed journal) devoted to trade unions and the environment provides an historical overview of the position of the FNV on the environment (Leenders and Boog, 1990). It identifies three stages of development, coinciding more or less with those identified by Winsemius with regard to national policy. Until 1970 the traditional workers' movement (as Leenders and Boog call it) definitely considered nature secondary to economic development. Nature was considered to play an important role in the development of the working class, and so any environmental concern originated from the point of view of preserving nature.

The second phase, typified by the report by the Club of Rome and the first oil crisis of 1973, reached its climax around 1974. Environmental pollution and the negative aspects of economic growth came into view for the first

---

[6] This refers to the eight elements comprising the standard environmental protection sysytem (see also paragraph 2.3. of this report) a written environmental policy statement the intergration of environmental protection into the operation of the company an environmental programme a measurement and recording programme an internal control system internal and external reporting internal information and training; and a regular audit of the company's environmental protection system.

time, but they were considered to be primarily the responsibility of government and employers. The prevailing view was that environment and employment were contradictory. In general the unions did not adopt a very strong position on the environment and their concern was weakened considerably with the arrival of recession and a sharp rise in unemployment (in the early 1980s). The demands of the economy were once again considered to be primary, well above those of the environment.

The third phase began in 1987. Leenders and Boog note the publication of a highly critical article by van de Biggelaar, then the FNV official responsible for environmental issues and now the director of the Nature and Environment Foundation, one of the most influential environmental organisations, operating independently of government and the social partners but occupying many important formal and informal positions (Van de Biggelaar, 1987). Later that year the FNV published, in cooperation with the most important environmental organisations, the memorandum "Investing in the Environment" (LMO/FNV, 1987).

In June 1988 it published "Trade-union Basis for Environmental Policy", whichtook a very serious look at environmental problems. It proposed a two-track strategy: on the one hand to continue to work on specific campaigns such asdeveloping cleaner products and production processes (at company level) and on campaigns on traffic management, transport and energy (at branch and national level); and on the other hand to unfold new activities in extending the old emancipatory endeavours, since the debate on environmental issues can affect key social values like democracy, solidarity and individual freedom.

The FNV's executive council accepted the proposals contained in the memorandum. A number of recommendations and commitments were made, which according to Leenders and Boog give a clear insight into the FNV's environmental policy. They are summarised as follows (Leenders and Boog, 1990, p. 88):

- A more stringent environmental policy will be adopted, and in principle all policy instruments should be used to this effect, even measures to reduce volumes;
- The unions should become involved in an advisory capacity in the development of national environmental policy;
- Affiliated unions should engage in regular environmental consultation with FNV headquarters and with environmental experts from outside the union movement;
- The FNV will join the Centre of Energy Conservation and the Nature and Environment Foundation;
- A think-tank should be set up between the unions, environmental organisations and the universities;
- Every trade union should appoint someone with responsibility for environmental issues;
- The FNV will explore the possibilities of an agreement with employers'organisations on the environment;

- The FNV will organise a consciousness-raising campaign among its members;
- The union's Institute for Technology Advice will be geared to organising environmental activities;
- Consideration will be given to topics which should be brought into collective bargaining (such as refusal to working with pollutants);
- Activities should be coordinated and fine-tuned at regional and local level;
- The FNV will have consultations with the Ministry for the Environment and the Ministry of Economic Affairs with regard to financing activities;
- Results will be communicated to European partners and
- Special attention will be paid to a clear presentation to members.

In another memorandum published in 1989 the FNV called for "a stricter environmental policy" and dealt with its implications for the union (FNV, 1989a). These papers were followed by a discussion project with the members, entitled "the signal on green" (FNV, 1989b). Contacts with the two employers' organisations resulted in the agreement between the FNV and VNO mentioned earlier. Further development of the planned activities are slowly bearing fruit, although some activities are stagnating.

Some FNV member unions have launched their own initiatives. The FNV Graphical Union published a booklet entitled "Clean: your job too" (Druk en Papier FNV, 1989). The FNV Manufacturing Union produced a publication on the union and the environment and on company environmental protection systems (Industriebond FNV, 1991a, 1991b). And the FNV Centre for Works Councils published a booklet on environmental protection in offices (Van de Schaft and van den Nieuwenhof, 1990).

In 1989 the CNV Manufacturing and Food Industry Union published an environmental policy plan (Industrie- en Voedingsbond CNV, 1989). As mentioned earlier, this lays great emphasis on the instrument of collective bargaining and raising the environmental awareness of union members. It also gives special attention to polluting industries which will be forced into closure in order to protect the environment. It proposes the creation of a support fund for those who lose their jobs in the process.

Several union and commercial institutes have developed courses on the environment. The CNV's educational institute provides a course on co-determination and the environment (Slotemaker - de Bruïne Instituut, 1989; de Kam, 1991). One of the activities financed within the framework of the promotional programme "Environmental Protection in Companies" is the development of further courses and training activities for workers involved in co-determination on environmental issues. Further material from the educational institutes of both the FNV and the CNV will therefore become available in the near future.

The unions, in particular the FNV, also took part in the discussions on company environmental protection systems. It is noticeable that increasingly common cause is made with environmental organisations, and that continuing differences are merely a matter of emphasis rather than principle. When the employers' organisations unveiled their concept of company environmental protection systems, the reaction from the environmental organisations was on the whole positive (Berends and Mol, 1988; Sprengers, 1989), although they criticised the employers' emphasis on a reduction of government inspection and their reluctance to make data on company environmental performance available to the public.

In their response the environmental organisations made three proposals. They called for compulsory public environmental reporting, whereby companies would be obliged to issue an environmental report in accordance with legally defined standards, auditing of reports by an environmental accountant, and the possibility of compulsory external environmental auditing. The unions reacted equally positively to the concept developed by the employers' organisations, although they tended to lay greater stress on the issue of workers' involvement in environmental protection.[7]

Both unions and environmental organisations on the whole reacted positively to the government's "Company Environmental Protection". The FNV called it "an important contribution" (FNV, 1990, p. 1). The Nature and Environment Foundation commented "better late than never" (Stichting Natuur en Milieu, 1989, p. 2). Both sides reiterated their calls for compulsory reporting and stressed the need for a compulsory external environmental audit. The environmental organisations also stressed the importance of government inspection. The FNV argued that insufficient attention was being paid to the relationship between environment and working conditions. It suspected that the Ministry of Social Affairs and Employment had not been properly involved in the preparation of the memorandum.

In conclusion it can be said that the government and workers' and environmental organisations responded positively to the willingness expressed by employers to behave more responsibly towards the environment by means of the introduction of company environmental protection systems. The issues of whether company data on environmental performance should be made public and whether government inspection is required are still under discussion. Yet another question concerns the involvement of workers. This is a matter which deserves more detailed analysis (see Le Blansch 1990, 1991a).

---

[7] In an environmental organisations publication at this time, Berendsand Mol (1988,p.17) noted a tendency, especially within the FNV Manufacturing Union, to accept the employers position, although at the same time they observed a more active attitude the environmental arising from their concern with employment.

4.4 Programmatical statements at company level

4.4.1 The advocates

Most policy papers on company environmental protection touch on the desirability, if not the necessity, of involvement from the side of employees.[8] The government has never let any doubt arise over the importance it attaches to this. In the NEPP it declared, "In the development and implementation of environmental protection systems the involvement of employees is an important factor, both at national, branch and company level" (Tweede Kamer, 1988-89, 21137, 1-2, p. 161.). Why this should be important is not explained. In "Company Environmental Protection" the government also expressed the view that it is self-evident "that employees or their representatives ... are involved in the creation and implementation of environmental protection systems" (Tweede Kamer, 1988-89, 206333, 2-3, p. 33f.). Yet again, there is no explanation why this is considered desirable.

The CNV Manufacturing and Food Industry Union is on record as saying, "The works council should play a role in the creation, organisation, implementation and control of the various aspects of company environmental protection" (Industrie- en Voedingsbond CNV, 1989, p. 22). The FNV also attaches importance to workers' involvement. In its response to the government's memorandum it declared that the involvement of workers is such an essential condition for the success of environmental protection systems that this should be made a criterion in the interim evaluation process to be undertaken in 1992 (FNV, 1990, p. 1). Here again no reason is given for this view.

Why this involvement is desirable has been dealt with in a range of other publications and from a range of different standpoints. A comparison of the various arguments yields a rough division into two main types:[9]

- Workers and their representatives must be involved in company environmental protection because this is in their interest; this may be called the "worker's perspective"; and

---

[8] The concept of involvement and co-determination are closely linked. Involvement is used in this context in the sense of participation co-determinations one of its possible manifestations. for an eleaboration of these concepts this report will rely on the literature on co-determination. no further definitions of terms will be attempted here.

[9] Compare Gevers, 1982. With regard to health and safety he distinguishes between practical and mormative motives for regulating workers control. Within the context of co-determination on environmental policy, practical motives (such as for environmentally friendly behaviour maintenance and developemnt of environmental measures) would be better described as motivation from an environmental perspective. Normative grounds in turn correspond to motivation from workers perspective.

- Workers and their representatives must be involved in company environmental protection because this is in the interest of a clean environment; this may be called the "environmental perspective".

A third type of argument is used considerably less frequently, and less publicly, but is of great influence especially at company level. This type of argument concerns the interest of the organisation's operation, and can be called "the functional perspective".

The different arguments will be analysed on the basis of this division.

4.4.2 Arguments from the workers' perspective

From the workers' perspective, the following arguments have been put forward to justify the involvement of workers in company environmental protection.

(1) Because environmental protection touches fundamentally on their safety and health, workers should be able to co-determine it.

It was noted earlier that the internal and external environment can be closely interrelated. Zwetsloot (1989), for instance, describes working conditions and environment as "twins" and distinguishes six possible types of relationship, both positive and negative and direct and indirect. [10] Several advocates argue that wherever protection of the environment and protection of working conditions overlap, workers should have a say. The National Environmental Forum (LMO), an informal grouping which brings together environmental and consumer organisations and trade unions, follows this line when it justifies involvement of workers in environmental protection on the grounds that "they after all experience on a daily basis the problems and dangers of the internal environment" (Sprengers, 1989a, p. 16).

So too Sprengers, when he draws the attention of works councils to the fact that environmental protection measures may lead to an improvement or a deterioration in working conditions (Sprengers, 1989b). According to Gevers (1982, p. 58) normative grounds can be sited to substantiate this argument, which follow on from the right of involvement in the protection of physical integrity and in the protection of health at work. These are the same normative grounds which in all member states of the European Community have led to the regulation of powers on health and safety (such as in the Working Environment Act in the Netherlands).

---

[10] He names the following twin relationships (a) environmental and working-environment problems which can be traced to the same roots, (b) environmental problems which may cause a working environment problem and (c) vice versa (d) environmental and working environment problems which reinforce each other (e) parallel working environmental problems (wetsloot, 1989,p308)

(2) Because the workers are co-responsible for the pollution caused by the company, they should have co-determination on its environmental policy.

This line of argument has been used, for instance, in the environmental policy plan of the CNV Manufacturing and Food Industry Union: "In the end what matters is that we are conscious of the fact that the jointly shared responsibility of employers and employees for cleaning the general environment and keeping it clean will occupy us for a very long time to come" (Industrie- en Voedingsbond CNV, 1989, p. 5). In the same text reference is also made to the Christian mission to "dress and keep" the earth as mentioned in Genesis 2:15 (p. 15).

Sprengers (1989b) gives the fact that employees are also held co-responsible outside the company as a justification for co-determination. The FNV (1990, p.8) points out both the employer and employees are liable for violations specified in a number of environmental laws. This is perhaps a reference to the Environmentally Hazardous Substances Act and Soil Protection Act. For this reason the FNV argues for "identification of and information on these laws and other regulations" and for "education, instruction and so on".

(3) Because decisions related to environmental protection affect workers directly, they should co-determine them.

Gevers (1988) places this principle within the context of the striving for a humanisation of labour.[11] He quotes the government memorandum at the adoption of the Working Environment Act, in which the wellbeing of the workers is directly linked to the degree to which they control their own work situation, "Wellbeing in connection with labour is concerned with the scope offered by the conditions, organisation and content of work for taking responsibility, input and creativity by the workers" (Tweede Kamer, 1978-79, 14497, 5, p. 7).

Decisions on environmental policy affect workers if for no other reason than the issue of continuity of employment. A report by the Environmental Education Foundation (SME) and the Centre for Energy Conservation and Clean Technology (CE) follows this argument and gives as a reason for workers' involvement the fact that "a company which pursues bad environmental management risks a government decision at some point to suspend its operations or close down it down altogether. This puts the issue of employment centre stage. For the workers and the works council it is therefore imperative to ensure that the management takes appropriate environmental measures" (SME/CE, 1988, p. 13).

---

[11] For Geers (1986,p11f) the concept of the humanisation of labour has three components (a) safety, health and hygiene, (b) a voice in the management of the company, and (c) responsibility, development and creativity

As another reason for the involvement of the works council in environmental protection, Sprengers (1989b) mentions the possible changes in tasks and responsibilities which could result from it.

4.4.3 Arguments from the environmental perspective

From the environmental perspective, the following arguments have been put forward to justify the involvement of workers in company environmental
protection.

(1) The workers should be involved in environmental protection because this motivates them to improve their environmental behaviour.

Gevers concludes that "if the freedom and individual responsibility are taken away by excessive regimentation, then an important incentive for an active commitment to the safety of oneself and of others disappears. Conversely, an extension of responsibility leads to a greater commitment" (Gevers, 1982, p. 52). In this line a recent VNO/NCW publication, "Environmental Reporting by Companies" argues for the provision of information to employees: "The cooperation of the individual employee is essential to the success of the environmental protection system. By increasing understanding [of environmental pollution and protection] individual awareness and the incentive to behave in an environmentally responsible way will increase" (VNO/NCW, 1990, p. 3f.).

Environmental behaviour outside the workplace can also improve through involvement in environmental protection within the company. In its policy plan the CNV Manufacturing and Food Industry Union identified a number of roles in which CNV members deal with the environment: as union members, as citizens, as car users, consumers and tourist. The CNV sets itself the task of initiating a process of consciousness-raising (CNV, 1989, p. 17). One of the arguments in favour of workers'. involvement put forward by Ket (1989, p.1) is that employees are also consumers and therefore an important factor in environmental policy.

(2) Workers should have co-determination on environmental protection because they can fulfil a useful role in enforcing environmental legislation or where appropriate remind employers of their responsibilities concerning the environment.

This line of argument can be found in many quarters. An employee at the Hoogovens steelmill is quoted in the Praktijkblad voor Medezeggenschap as saying that the management "is far more sensitive to criticism from within than from outside, from environmental action groups, for instance" (Anon., 1988). In an interview with OR-Informatie, a magazine on co-determination (Verbakel, 1989), Ed Nijpels, then Minister for the Environment, showed himself a strong supporter of corporate "environmental accountancy" monitored by the works council.

At a congress of the CNV Wood and Building Union he expressed himself in similar terms, and did not exclude statutory measures if it emerged that the council did not have sufficient monitoring powers (Ket, 1989, p. 6). And in a preface to a publication by the environmentalist movement, "Environmental Protection in Companies, Why It Matters" the environmental science lecturer Hommes, commenting on the call by the authors for a mandatory external public audit, argues that they underestimate the internal monitoring within a company (Berends and Mol, 1988, p. 5).

(3) Workers should be involved in environmental protection because they have expertise and experience which may be useful in the development and application of environmental measures.

Although this argument cuts more ice in the debate on worker involvement in measures concerned with increasing subjective wellbeing, it is also used in the context of environmental protection. Thus the FNV gives as another reason for the involvement of workers that "experiences in the sphere of traditional working conditions can be very relevant" (FNV, 1988, p. 5). The SME/CE paper also adopts this position, "Workers constantly come into direct contact with environmental problems: leaking pipes, saturated filters, used oil which gets into sewers, and so on. Therefore most decisions on intervention will have to be taken on the shopfloor" (SME/CE, 1988, p. 13).

The Research Council on the Environment and Nature (RMNO) recommends that environmental protection should be given shape not only in a top-down but also and in particular in a bottom-up approach, "In addition, the actual implementation of environmental protection systems and their incorporation into general corporate policy are of great importance. This is a task for line management and employees. That is why it is important that the development of policy takes account of implementation problems which may occur at the lower levels of an organisation, the so-called "bottom-up" approach" (RMNO, 1989, p.14). Gevers puts the potential contribution of workers also in the light of selective perception, applying to both management and workforce, and links to this the condition that for an optimum input from this self-perception of workers, possibly inadequate knowledge should be supplemented with the help of training or external expertise (Gevers, 1982, pp. 56f.)

4.4.4 Arguments from the functional perspective

Employers and their organisations tend to approach workers' involvement primarily from a functional perspective. As this concerns in the first place a specific company interest and less a collective public interest, the arguments from this perspective are less often stressed in public as well.

Still some of the arguments sited in the sections above clearly originate from a functional perspective, like those concerning involvement in order to motivate personnel to behave in accordance with company environmental policy. The same goes for arguments stressing the importance of expertise and knowledge on the shopfloor to become available for management (the so-called "bottom-up" approach).

VNO/NCW further elaborate these arguments, also concerning the workforces' commitment to the company environmental protection policy. They emphasise the importance of "close involvement of all personnel (as) a precondition for the environmental protection system to function properly" (VNO/NCW, 1986, 21). And the supply of information to employees is (also) argued as follows, "A greater understanding will lead to a motivated staff and a work force that is motivated to take the responsibility of environmental hygiene seriously. Furthermore, it is of the utmost importance that employees are well informed to enable them to react correctly when remarks concerning the company environmental policy are made by the general public" (VNO/NCW, 1991, 5).

Sometimes arguments from this perspective are stated in an indirect manner. This concerns for instance a derivative of the "quality of working life" argument. Thus, the government's "Company Environmental Protection" mentions "contributing to the quality of the production process and products" and "positive influence on staff motivation", among others, as interests of industry in environmental protection (Tweede Kamer, 1988-89, 20633, 2-3, p. 11). Also the report by the Research Council on the Environment and Nature (RMNO) (RMNO, 1989, p. 14), mentions the view of the quality of labour and safety and environment as primary production factors rather than residual matters, as environmentally relevant developments.

If not so very often used in public, these arguments are often encountered in case studies concerning workers' involvement in company environmental practices. For example in one Dutch case study (Le Blansch, 1991b) an environmental coordinator is cited, "It is the task of management to provide for policy, means and opportunities. If these fall short then management is pleased to be informed of this. Problems may arise in implementation and in this management an works council may well cooperate".

This quotation demonstrates the limits of the functional arguments as does, even more so, a statement on workers involvement by an environmental coordinator from another company: "It helps integrating the protection of the environment in the company's activities. Co determination however must remain functional, 13,500 small decision-makers are of no use".

In short, three kinds of functional arguments can be distinguished:

1) Workers should be involved with company environmental protection because this motivates them to behave in accordance with company environmental policy (and the interest of the company in general).
2) Workers should be involved in the formulation of company enivronmental policy because they have expertise and experience which may be useful in the development and application of environmental measures.
3) Workers should be involved with company environmental protection because they contribute to the organisation's outward image.

4.5 Discussion

The various arguments for workers' involvement have been outlined above. They suggest, more or less, that the workers', company's and environmental interests are closely connected and interrelated. Again, some qualifying remarks need to be made at this point. It has already been pointed out that involvement and co-determination may be quite difficult to realise, both from a legal and political power point of view. Therefore it is by no means certain that all workers are in a position to exercise influence.

Even when workers have the opportunity of exercising influence over company policy, it is by no means certain that they will use it to protect the environment. A prerequisite is that they are concerned about the environment, that they are aware of the environmental pollution caused by the company, that they feel partly responsible for it, and that they have a degree of idealism about environmental issues. In short, a particular attitude is required.

Nelissen (1988) has made a study of the internalisation of environmental norms from a socialisation angle. As environmental standards become a more elementary part of the dominant cultural pattern, environmentally friendly behaviour becomes a question of conforming to normal patterns of behaviour. He therefore adopts a "socialisation" approach to the environmental question. He identifies a number of barriers to effective socialisation. Although he focuses primarily on those mainly responsible, the employers, most of the barriers he mentions can also be an obstacle to workers' involvement in environmental protection. What springs to mind here is the influence of the social environment, the absence of alternatives, or relative ignorance of environmental issues. An important threshold identified by Nelissen is the so-called "cost-benefit barrier". This arises when a positive attitude to the environment costs the workers more than they gain from it. The existence of this barrier presupposes a difference between the workers' interests and environmental interests.

Possible differences between the workers' interests and environmental interests can be discussed either from a macro (or collective) point of view, or from a micro (or particular point) of view. Viewed at the macroeconomic level, it is open to question as to whether any such difference exists. For the quality of the life of the workers will be directly linked to the overall quality of the environment in which they find themselves. And a company which consistently neglects its critical dependence on the environment will eventually go under. But it is certainly true that at the microeconomic level, within the individual company, tensions between the interests of workers and the environment can be observed, quite apart from the question as to whether these interests are correctly perceived. This is why involvement of workers does not always lead to the protection of environmental interests. A publication by the FNV, "Trade Union Basis for Environmental Policy" (1988) identifies four types of friction between workers' and environmental interests which may be the cause of this:

(a) term effects: effects on employment tend to reveal themselves in the short term, while environmental effects emerge in the long term;

(b) cost allocation: employment arguments can sometimes overcome the "polluter pays" principle;

(c) cumulative effects: it is often difficult to identify individual transgressors, since the damage done is the sum total of many forms of relatively "harmless" environmental pollution and

(d) interest allocation: it is difficult to demarcate the general interest from the interests of those directly involved.

The upshot of all this is that at the micro level and in the short term conflicts of interest between workers' and the environment can occur in specific cases. Reijnders (1990) and Leisink (1989) provide a short survey of cases in the past where these conflicts have come to the fore in the form of contradictory statements by proponents of workers' and environmental interests at macro and micro level. These are summarised below:

- Coal-fired power station in Amsterdam: FNV-Amsterdam in favour, the environmentalist movement and FNV headquarters against;

- Phosphate-free washing powders: the environmental movement and FNV headquarters in favour, the union at Hoechst against;

- Banning of Dinoseb (a chemical company): environmental movement in favour, trade union against;

- Reclamation of Markerwaard: several unions in favour, environmental organisations against and

- Abolition of travel costs subsidies: environmental movement in favour, union against.

Of course the changing times have not passed either the workers' organisations or the environmentalist movement by. Workers' organisations have recognised the contradictions mentioned above and have tried to resolve them. Partly in the light of its current "success" the environmental movement has come to better understand the interests of workers' organisations, and there have been growing calls for joint action. Nevertheless, in specific cases some painful choices will still have to be made. Under certain circumstances it will be unrealistic to expect workers and their representatives and environmental organisations to take an overly balanced view of the primary interests which they seek to represent.

5. SUMMARY AND RECOMMENDATIONS

In the previous chapters an overview is given of the different ways in which industrial relations and the environment are interconnected in the Netherlands. Firstly, Dutch legal conditions under which industrial relations affect or may affect environmental issues and vice versa have been presented. On the one hand this concerns environmental legislation in general and in particular the development towards (government sponsored) self-regulation by means of company environmental protection systems. On the other hand it concerns Dutch legislation on the working environment, particularly the Working Environment Act (WEA). This is primarily an enabling act, defining the role of the parties involved and their respective rights and duties. It can be noted that the legal means for co-determination are slowly being extended towards the environmental realm. However some qualifying remarks have been made about both the uncertainty of actual workers' involvement in environmental protection at company level, and the still limited extent to which co-determination rights can be applied to environmental issues.

Next, voluntary agreements at the national, branch and company level have been reviewed. At national level existing structures like the Social Economic Council have been used for consultation on environmental issues. As a result of a national agreement between social partners, special structures have been created, in which, however, negotiations seem to have come to a deadlock. A dominant power perspective on assessing one anothers' motives and differences in style concerning the approach of environmental issues, appear to present obstacles for the further development of a common approach on environmental problems by social partners.

At branch and company level collective bargaining, which is among the principle areas in which union policies on the environment are being developed, has led to a growing amount of collective agreements on the environment. At company level workers' involvement is definitely increasing, although valid data on the exact contents and intensity of involvement are difficult to find.

The starting points of involvement and co-determination practices appear to be working conditions and everyday practice, whereas the agreements seem to so far present more a codification of practices than a modification. However it is far too early for a comprehensive evaluation.

Finally, the programmatic statements, demands and campaigns, both from employers' and workers' side, have been described. From employers' side, the most important issue at stake has been the introduction of Company Environmental Protection Systems and its acceptance by government as an alternative to direct regulation (from 1986). Here the employers' environmental bureau, the BMRO, played a proactive and successful role. The result so far can however be considered to be rather paradoxical, as the postponement of direct regulation legitimates stronger public attention to what is actually happening inside companies.

The unions also show a "greening" tendency, with the major environmental policy push dating from 1987. It is noticeable that in taking position in public increasingly common cause is made with environmental organisations. A special point of interest from the union side (and in part from the side of the employers' organisations, government and environmental organisations too) concerns the involvement and co-determination of workers in company environmental policy. The FNV is even arguing that this should be made a criterion for the evaluation of the state of corporate environmental policies in 1992.

In part these cases are argued from a workers' perspective. What matters from this perspective is the influence of environmental protection on health and safety, the general work situation of the worker, or the aspects for which the worker is partly responsible. From an environmental perspective involvement of workers is advocated because it would have a motivating effect on their environmental behaviour, would enable them to play a useful role in supervising the environmental behaviour of the employer and because they have valuable knowledge and experience. From a functional perspective workers' involvement is advocated because it motivates them to behave in accordance with company environmental policy, because it enables knowledge and expertise at the shopfloor to become available for management and because workers contribute to the organisation's outward image.

Practical co-determination based on the coincidence of the workers', the company's and environmental interests applies above all at the macroeconomic level. At the micro level, frictions can occur. Those mentioned here are term effects, cost allocations, cumulative effects and interest allocation. In cases where these conflicts arise the involvement of workers may have a counterproductive effect in environmental terms.

Problematic co-determination, possibly insufficient socialisation and conflict between workers' and environmental interests may stand in the way of a meaningful workers' involvement in the development of company environmental policy. So far too little is known about the extent and form of this involvement and more research in this area is needed.

On the one hand the question arises: under which conditions may treatment of environmental issues within industrial relations serve the interests of the environment? When the implementation of company environmental policy becomes the aim and result of consultation and negotiation between employers and employees at all levels, then factors such as the overall economic situation and balance of power in companies and branches may start playing an overriding and improper role in the decisions on the collective environment. It is by no means certain that workers' influence will be strongest in the branches which generate the most pollution. Indeed, in the older branches of industry, the heaviest polluters, it may be asking rather a lot of workers to protect the interests of the environment when this may mean a real threat to their own jobs. More research is also needed here, particularly in relation to the possible need for a branch-wide intensification of union and training activities, if not retraining activities.

But the question also arises: under which circumstances may workers' interests be served by a general argument for involvement of workers and their representatives in environmental issues? Apart from the possible occurrence of friction between workers' and environmental interests discussed earlier, there is also the question of how this argument relates to earlier and other calls and claims in the areas of disarmament, apartheid, women's rights or the four-day working-week.

Seen in this light, an inability to give shape to an effective involvement of workers may quickly lead to an erosion of the credibility of co-determination as an (ideological) institution.

Related to this is the question as to whether and how the activities of the works council can be structured in such a way that attention to environmental protection is not just the umpteenth task put on the shoulders of a handful of committed employees in the company, but which cannot be realised. Here too more research is in order. In the worst possible case the argument for worker involvement can be turned into its opposite and lead to the workers being assigned co-responsibility for a situation only partly, if at all, of their own making which may lead to an exceedingly messy entanglement of interests. This would neither be in the interest of the workers nor of the environment.

REFERENCES

Anonymus (1988): 'Het dilemma: milieuzorg en werkgelegenheid'. Praktijkblad voor medezeggenschap, February 1988.

Bayens G. and R. Prins (1990): Labour inspectorate and the quality of working life in the Netherlands. Voorburg, DGA.

Berends, Wilma and Tuur Mol (1988): Milieuzorg in bedrijven, ons een zorg. De visie van Milieudefensie en Natuur en Milieu op interne milieuzorg, milieu-accountancy en milieu-auditing. Amsterdam.

Berg, D. van den and A.J. de Raaff (1990): Op weg naar bedrijfs milieu diensten. 's Hertogenbosch: Nehem.

Biggelaar, A. van de (1987): 'De vakbeweging en het milieu' Natuur en Milieu, March 1987, pp. 4-50.

Bloemers, F.W. and H. Kraaij (1987): 'Vertrouwen is goed maar controle is beter'. Milieu en Recht, 1987, 3, pp. 87-93.

Calkoen, P. and K. ten Have (1991): Bedrijfsinterne milieuzorgsystemen. Tilburg: IVA, instituut voor sociaal wetenschappelijk onderzoek.

Centrum voor Energiebesparing en schone technologie, Stichting Milieu-Educatie (1988): Bedrijf en milieu. Utrecht: CE/SME.

COB-SER (1991): Handboek Woon-werkverkeer (+ Wegwijzer). Den Haag.

Commissie Bedrijfsinterne Milieuzorgsystemen (1988): Milieuzorg in samenspel. 's Gravenhage.

Dienst Collectieve Arbeidsvoorwaarden (1991): CAO regelingen 1991. Den Haag.

Doorewaard, M.E.M. (1990): Milieuwetgeving en het bedrijfsleven; de paradoxale rol van belangenorganisaties. Groningen: Wolters Noordhoff.

Druk en Papier FNV (1989): Schoon ook jouw werk; Vakbondsleden aan de slag voor een schoner bedrijf en een schone omgeving.

Federatie Nederlandse Vakbeweging (1988): Een vakbondsbasis voor milieubeleid. Amsterdam.

Federatie Nederlandse Vakbeweging (1989a): 1989, Omslag naar een scherper milieubeleid, ook voor de FNV. Amsterdam.

Federatie Nederlandse Vakbeweging (1989b): Material for a discussion-project, Het sein op groen. Amsterdam.

Federatie Nederlandse Vakbeweging (1990): 'Schriftelijke reaktie op de notitie Bedrijfs-interne milieuzorg': Amsterdam.

Geers, mr. A.J.C.M. (1988): Recht en humanisering van de arbeid. Monografieën Sociaal Recht, nr. 6. Deventer: Kluwer.

Gevers, J.K.M (1982): Zeggenschap van werknemers inzake gezondheid en veiligheid in bedrijven; de rechtsontwikkeling in de lidstaten van de
Europese Gemeenschap. Deventer: Kluwer.

Hengelaar, Gerjos (1989): De ondernemingsraad en het milieu. Utrecht: Stichting Milieu-Educatie, Utrecht.

Hofstra, N., J. Timár, H. Verdonk and F. de Walle (1990): Milieubedrijfsvoering; problemen en perspectieven; een vergelijking tussen Nederland en Californië. Rotterdam: 'le manageur'/SCMO-TNO/EUR-BKE.

Industriebond FNV (1989): Onderzoek Chemie 1989; Totaaloverzicht resultaten. Over werkgelegenheid, scholing, beloning, beoordeling, veiligheid, gezondheid en milieu. Rotterdam: IB-FNV.

Industriebond FNV (1991a): De industriebond FNV en het milieu. Amsterdam IB- FNV.

Industriebond FNV (1991b): Milieuzorgsystemen en de Inustriebond FNV. Amsterdam IB-FNV.

Industrie- en voedingsbond CNV (1989): Voor een schoner milieu, milieubeleidsplan van de industrie- en voedingsbond CNV. Nieuwegein.

Industrie- en voedingsbond CNV (1990): 'Stand van zaken milieubeleid I.V.B. CNV per mei 1990'. Nieuwegein.

Kam, E. de (ed.) (1991): Milieu en Medezeggenschap. Doorn, Slotemaker-de Bruïne Instituut.

Ket, Petra (1989): 'FNV en het milieu'. (Supervisor Gert Spaargaren) Amsterdam.

Klatten, Richard (1989): Address on the FNV conferences on works council and environmental protection.

Koninklijk Nederlands Ondernemers Verbond / AMRO-bank (1990): Milieu, een hele onderneming; Onderzoeksresultaten. Amsterdam.

Le Blansch, C.G. (1990): Werknemers en bedrijfsinterne milieuzorg; een vooronderzoeksverslag annex onderzoeksvoorstel. Utrecht: Coördinatiepunt Wetenschapswinkels Utrecht, RUU.

Le Blansch, C.G. (1991a): 'Betrokkenheid van werknemers bij bedrijfsinterne milieuzorg. Een verkenning van een nieuw terrein van medezeggenschap. Tijdschrift voor Arbeidsvraagstukken, vol. 7 (1991), nr. 1.

Le Blansch, C.G. (1991b): Werknemers en milieuzorg' (working title). To be published: autumn 1991.

Leenders, P. and B. Boog (1990): 'Van natuurbeschouwing naar milieubeleid; de trage vergroening van de FNV'. Tijdschrift voor Arbeid en Bewustzijn, 1990/2, pp. 85-90.

Leisink, P. (1989): Structurering van arbeidsverhoudingen; een vergelijkende studie van medezeggenschap in de grafische industrie en in het streekvervoer. Dissertation, Utrecht.

LMO / FNV (1987): Investeren in milieu. Milieuoverleg RCO / Vakcentrales (1989): Notitie inzake milieuzorg. Den Haag.

Ministerie VROM (1990): Circulaire betreffende criteria subsidietoekenning projecten bedrijfsinterne milieuzorg. Den Haag.

Ministerie VROM (1991): Stimuleringsprogramma Bedrijfsinterne milieuzorg. Den Haag.

Ministerie SoZaWe (1987): Jurisprudentie Medezeggenschap. 22 July 1987, nr. 87/2468. Den Haag.

Mintzberg, H. (1979): The structuring of organizations. Englewood Cliffs: Prentice Hall.

Mulder, K. and J. Benders (1991): 'Medezeggenschapsregelingen vergeleken en gewogen; een overzicht'. Pre-publication Wetenschapswinkel Rechten RUU. Used with permission.

Nationale Investeringsbank (1991): Onderneming en milieubeheer. Den Haag: NIB.

Nelissen, N. et.al. (1988): Het milieu: Vertrouw, maar weet wel wie je vertrouwt. Zeist: Kerckebosch.

Praktijkblad voor de Medezeggenschap (1990): 'VGW in profiel; Onderzoek Industriebond FNV'. Praktijkblad voor de medezeggenschap, 1990, p.22-23.

Peters, A.A.G. (1979): 'Recht als project'. Ars Aequi 28 (1979), 11, p.245/881-256/893.

Raad voor het Milieu- en Natuuronderzoek (1989): Milieumanagement bij bedrijven; van concept naar toepassing. Programmerings- en Studiegroep Milieu en economie; Werkgroep Bedrijfskundig milieu-onderzoek. Dr.

W.A. Hafkamp (chairman); Drs. O.J. van Gerwen (secretary); Rijswijk: April 1989.

Reijnders, L. (1990): 'Arbeid en Milieu; De ecologische rol van de vakbeweging'. Zeggenschap, tijdschrift voor vakbewegingsvraagstukken. January/February 1990, vol. 1, 1, pp. 46-49.

Schaft, M. van de and R. van den Nieuwenhof (1990): Schoonschrijven; milieuzorg in kantoororganisaties. FNV Centrum voor Ondernemingraden, Amsterdam.

Schot, J., B. de Laat, R. van der Meijden en H. Bosma (1990): Geven om de omgeving; milieugedrag van ondernemingen in de chemische industrie. STB-TNO, Apeldoorn.

Sociaal Economische Raad (SER) (1991): Milieu en economie. Den Haag.

Slotemaker-de Bruïne Instituut (1989): Milieu en Medezeggenschap. SBI, Doorn.

Sprengers, Piet (1989a): Ideeën voor milieuzorg. Wilma Berends ... et.al. (compilation); Peter Visser (ed.). Utrecht: Landelijk Milieu Overleg.

Sprengers, Piet (1989b): Address to the FNV conference on works councils and environmental protection.

Stichting Natuur en Milieu (also on behalf of Milieudefensie) (1989): 'Schriftelijke reaktie op de notitie Bedrijfsinterne Milieuzorg'. Utrecht.
Stichting Milieu-Educatie (1989): Report on research for the production of a video on workers and the environment. Utrecht: SME.

Teijlingen, Hans van (1989): 'Centrale rol OR bij milieuzorgsystemen'. OR- informatie vol. 7, 11 October 1989.

Teulings, A. (1981): Ondernemingsraadpolitiek in Nederland. Doctoral dissertation, Amsterdam.

Touche Ross Europe Services (1990): European Management Attitudes to Environmental Issues. Brussels.

Tweede Kamer (1988-1989), 21137, 1-2: Nationaal Milieubeleidsplan; Kiezen of verliezen. Also: National Environmental Policy Plan (NEPP).

Tweede Kamer (1988-1989), 20633, 2-3: Notitie bedrijfsinterne milieuzorg.

Tweede Kamer (1989-1990), 21137, 20-21: Nationaal Milieubeleidsplan-plus (NEPP+).

Vakbondsschool (1988): Vakbond & milieu. Ledenscholing FNV.

Verbakel, F. (1989): 'Ondernemingsraden moeten aan de bel kunnen trekken'. OR-informatie vol. 5 (1989) no. 5, pp. 4-8.

Verbond van Nederlandse Ondernemingen VNO / Nederlands Christelijk Werkgeversverbond NCW (1982): Visie van VNO en NCW op milieu en ruimtelijke ordening. Den Haag.

Verbond van Nederlandse Ondernemingen VNO / Nederlands Christelijk Werkgeversverbond NCW (1985): Het milieubeleid nader bekeken. Den Haag.

Verbond van Nederlandse Ondernemingen VNO / Nederlands Christelijk Werkgeversverbond NCW (1986): Milieuzorg in bedrijven. Den Haag. Also: Environmental Protection in Companies.

Verbond van Nederlandse Ondernemingen VNO / Nederlands Christelijk Werkgeversverbond NCW (1986): Onderneming en milieu. Den Haag.

VNO (1989): 'VNO-Coördinatienota CAO-onderhandelingen 1990'.

Verbond van Nederlandse Ondernemingen VNO / Nederlands Christelijk Werkgeversverbond NCW (1990): Milieurapportage door bedrijven. Den Haag. Also: Environmental Reporting by Companies (1991).

Verbond van Nederlandse Ondernemingen VNO / Nederlands Christelijk Werkgeversverbond NCW (1991): Bedrijfstakken en milieuzorgsystemen: stand van zaken. Den Haag.

Windmuller, J.P., C. de Galan and A.F. van Zweeden (1983): Arbeidsverhoudingen in Nederland (fourth edition). Utrecht/Antwerpen, Het Spectrum.

Winsemius, P. (1986): Gast in eigen huis; beschouwingen over milieumanagement. Alphen aan den Rijn: Samsom H.D. Tjeenk Willink.

Zwetsloot, G. (1989): 'Arbeidsomstandigheden en milieu, een tweeling?'. Maandblad voor arbeidsomstandigheden vol 65, 1989, nr. 5, pp. 308-312.

INDUSTRIAL RELATIONS AND THE ENVIRONMENT:

SPAIN.

by

Ernest Garcia (Coordinator)
Rafael Gadea
Ignacio Lerma
Maria Luisa Lopez
Alicia Marcos
Jose Maria Ramirez
Antonio Santos Ortega

TABLE OF CONTENTS:

Page

1. INTRODUCTION..126

2. THE LEGAL FRAMEWORK...................................126
2.1 Introduction
2.2 Environmental legislation.............................128
2.3 Legislation on the working environment...............130
2.4 Health Legislation...................................131
2.5 Summary..132

3. COLLECTIVE AGREEMENTS IN HEALTH AND ENVIRONMENTAL
   ISSUES...133

4. POLICY STATEMENTS, CAMPAIGNS AND DEMANDS BY UNIONS AND
   EMPLOYERS..142
4.1 Introduction
4.2 Employers' organisation
    4.2.1 Chamber of Commerce initiatives...............145
    4.2.2 Policy versus practice........................146
4.3 Union organisations..................................147
    4.3.1 Comisiones Obreras (Workers' Commissions.....149
    4.3.2 General Workers' Union (UGT).................151
    4.3.3 The Environment in the Priority Union Proposal
          (PSP)..152
4.4 Summary..153

5 ATTITUDES OF INDUSTRIAL RELATIONS ACTORS TOWARDS THE
   ENVIRONMENT..154
5.1 Introduction and methodological clarifications
5.2 Attitudes of employers towards the environment.......155
5.3 Union attitudes towards environmental issues.........162
5.4 Summary..166

6 SUMMARY AND CONCLUSIONS..............................167

   APPENDIX 1...172

INDEX OF TABLES:

TABLE 1: TREATMENT OF HEALTH AND ENVIRONMENT
IN COLLECTIVE AGREEMENTS
VALENCIA REGION AND CATALONIA 1987

TABLE 2: TREATMENT OF HEALTH AND ENVIRONMENT
IN COLLECTIVE AGREEMENTS.
STATE AND VALENCIA REGION 1991

TABLE 3: SUMMARY OF BUSINESS ATTITUDES
TOWARDS THE ENVIRONMENT

1. INTRODUCTION

The environment and industrial relations: a strange sounding combination within a Spanish context. In the public's perception, "environment" conjures up images of ecologists, wide open spaces and consumption. Of course, industry has something to do with all this, being a source of pollution; but references made here are to technology, materials, products. The co-operation - conflict dynamic between the social actors of industrial relations is considered apart. "Industrial relations", for its part, is considered to be simply to do with jobs and wages; and at most involves health and safety at work. The line where the two meet appears somewhat blurred.

This situation has given rise to numerous difficulties in the composition of this report. Firstly, in relation to obtaining significant information, reference to previous studies has proved difficult, even more difficult as far as published studies are concerned, and virtually impossible with respect to any with a sociological focus. Revision of environmental and occupational legislation has revealed a uniting of the two, both relevant to our aim but a lack of specific guidelines. The review of a wide sample of collective agreements (chapter 3), at state and regional level, indicates the relatively frequent presence of matters associated with the internal environment (ie health and safety), but the absence of references to the external environment. Programmes and claims of company and workers' organisations (chapter 4) show there is evidence, on both sides, of an environmental discussion which is emerging as much more than just embryonic; but these also demonstrate that the issue is not considered to be one which overlaps into their mutual social relations.

The attitude analysis (chapter 5) adds two more difficulties. Even the most open ideology is notably cynical, and preaches - at least as an objective - conciliation between protecting the environment and the continuation of the prevailing growth model.

This makes the idea tend towards minimising conflict, and therefore, reduces the frequency and intensity of expression in the argument. On the other hand, attitudes tend towards the paradoxical. The criteria of widespread environmentalism are easily accepted, unless they reduce profits (in the case of employers) and apart from when they affect wages or jobs in a negative way (in the case of workers).

Of course, the slightly confusing presence of the object of the study surrounded by structured public arguments (collective agreements, programmes, congressional statements etc.) does not mean to say that this is always the case. There is a substantial amount of clear industrial practices in this field. However, an explanation of this characteristic would have meant having to recourse to studying cases, a task which would have led us away from the general focus of this report but an indepth explanation of these would require case studies.

We have basically used four types of sources: legislation, collective agreements, documents from union and employers conferences and meetings and last but not least, published documents. The few existing and available surveys on environmental issues opinion have been examined and several consultations have taken place with people in significant posts in the field of industrial relations.

We end this introductory note by acknowledging the cooperation of many people and organisations. We are grateful to Inés Ayala and Winni Woischnik, from the Social Action Secretariat of U.G.T (General Workers Union), to the Valencia Regional Confederation of CC.OO. (Workers Commission), to Angel Cárcoba and Joaquín Nieto from the Confederal Department of Environment and Ecology at CC.OO.; to the 1st of May Foundation, to Rafael Luengo, from the CEOE Secretariat for the Environment, the Valencia Cámara de Comercio (Chamber of Commerce). All the persons mentioned have kindly answered our questions and have given us access to many significant documents. Also to the documents service of the General Environmental Secretariat from the Public Works Ministry, for all the facilities provided in order to consult their archives. Finally to Sarah Whyte, for her patient and careful work in preparing the English version of this report.

2. THE LEGAL FRAMEWORK

2.1 Introduction

The Spanish Constitution recognises in article 37.1 the right to collective agreements, entrusting the Law with the task of guaranteeing this right in a free and autonomous way. This constitutional mandate is instituted by Law 8/80 of 10th March in the Workers' Statute in which Title III (articles 82 to 92) regulates bargaining and collective agreements. The Statute recognises in article 82 its legal status as a regulation, as does article 37.1 of the Constitution which recognises the binding force of collective agreements "which signifies the recognition of them not only as a simple contract, but as a legal standard" (Sala, T; 1990).

At the same time, the content of agreements depends to some extent on the character of collective occupational bargaining, according to article 37.1 of the Constitution. This allows the inclusion in bargaining of all those issues affecting industrial relations. The only material limit to the content of bargaining is "with respect to and coordination of those other rights protected by the Constitution with the same or greater intensity as the right to collective bargaining" (Sala, T; 1990).

Both texts, Constitutional and Statutory, leave the way open for protagonists of the conflict to project their interests on regulated specifications in the agreement, within the scope of valid laws. On the other hand, the development of the environmental collective agreement is restricted by the non-existence of a general environmental law and the consideration of the environment only as "the internal environment within the factory".

This regulatory delay and restricted perspective are important since they cause difficulties in the regulation and delegation of duties and responsibilities to different Ministries (Public Works, General Environment Secretariat; Health and Employment), Public Organisations, Autonomous Communities[1] or Environment Agencies and Local Authorities. The end result is that the regulatory framework becomes more complex and the participation of the latter in prevention, correction, follow-up, intervention and negotiation in environmental issues is much more complicated.

Adhering to this restriction which links the environment with occupational health, limits the actions of employers' organisations and trades unions and takes away a great part of the content of the agreements. It is worthwhile pointing out that the "national occupational health legislation is taken from a model inherited from the previous system. It continues to be deemed as health and safety legislation, occupational as a consequence, and destined to eliminate points of conflict in companies" (Alfonso Mellado, C; 1989) and therefore stands out as a double restriction.

The constitutional mandate with regard to the environment "the right of citizens to enjoy an environment suitable to personal development" (article 45 C.E.) and occupational health which obliges public powers to guard "health and safety at work" (article 40.2.C.E.) within the "general right to health protection (article 43.1) and the right to life and physical and
moral wellbeing" (article 15 C.E.) is developed by means of three regulatory channels occupational, health, and environmental.

These three regulatory frameworks present aspects which from their fields of duty allow employers' organisations and trades unions to project a global environmental consideration onto the content of collective agreements, as obviously certain responsibilities should be extended or specified.

---

[1] The Spanish State is composed of 17 Autonomous Communities, each comprising the former province or regions according to cultural identity. For example, the Autonomous Community of Catalonia is made up of 4 provinces (Barcelona, Girona, Lleida and Tarragona). Each Autonomous Community has its own Parliament and Government with administrative powers covering its whole area but with responsibilities which vary from one community to the next. Certain communities possess legislative powers, apart from those who depend exclusively on the Spanish State.

2.2 Environmental legislation

The establishment of the Spanish Constitution in 1978 contains a number of essential issues as far as environmental legislation are concerned:

a) The "constitutionalisation" of the right of citizens to "enjoy an environment suitable for personal development", as well as the duty of preserving it (article 45 of the Constitutional Text).

b) The obligation of public authorities from the date of the constitutional order onwards "to guard the rational utilisation of natural resources, with the aim of protecting and improving quality of life and defending and restoring the environment with the support of indispensable collective solidarity". This obligation involves, amongst other things, the State exercising a policing role in environmental matters which includes the power to impose administrative and penal sanctions.

c) The inclusion of this order in Chapter 3 of Title 1 of the Spanish Constitution "Guiding Principles of Social and Economic Policy", implies that, as this right is not considered fundamental (as it would be if it was included in Section 1 Chapter 2 Title 1: "Fundamental Rights and Public Liberties") it cannot be directly invoked but (article 53.3 of the Constitution) "must be preceded by action taken by means of positive legislation, legal practice, and public powers"[2].

d) The introduction into the political-administrative system of the Autonomous Community and, above all, the possibility that these Communities take on certain responsibilities in environmental matters.

Effectively article 149.1.23 establishes that the state has exclusive responsibility over the "basic legislation on environmental protection, without the danger of Autonomous Communities establishing additional standards. For its part, article 148.1., in reference to other matters where Autonomous Communities can assume responsibilities, includes, "management in matters of environmental protection" (article 148.1.9). Autonomous Communities have started developing laws and carrying out basic State legislation.

e) In relation to section b of article 45 of the Constitution referring to the possibility of imposing penal sanctions for failure to comply with environmental standards, it is essential to mention the nature law 8/86 of 25th June, which adds a new article to the Penal Code: 347 bis. This punishes "anyone who contravenes laws or environmental regulations, causes emissions or spillages of any type directly or indirectly into air, land, fresh or sea water, places peoples' health in serious danger or seriously endangers animal, wood, natural species or plant life".

---

[2] An example of this is that no one can demand in court that, on the basis of Article 45, a green zone be protected. the procedure is that legislative powers create a law which protects the green zone.

By means of this inclusion a new type of offence is created, "the ecological offence", which according to the above article, is punished with up to six months imprisonment and fines from between 175.000 to 5 million pesetas.

Sentences are more severe where an industry is operating secretly, an order from the enforcing authority to correct or suspend a polluting activity has been disobeyed, if false documents have been presented regarding environmental aspects or if inspection has been obstructed. All these reasons could result in the temporary or permanent closure of the company and the Courts would propose that the local administration intervene in the company with the aim of safeguarding workers'rights. Penal and legal doctrine define the environment as "maintenance of properties... and the developing conditions of species, in such a way that the eco-system maintains its subordinate systems and does not suffer harmful changes".
f) Local authorities (in particular Town Halls) are responsible for granting and supervising licenses for companies and industries, whose activities acquire "graded" status, that is to say those of unhealthy, harmful or a dangerous nature.

Of all the above, it is worth positively assessing the innovations which the Constitutional Text has brought about, one of which is the inclusion of "Ecological Offence". But it is paradoxical that responsibilities are dispersed, increasing the complexity of the regulatory system on environmental matters.

A number of environmental regulations are analysed below:

Environmental Impact Regulation. This regulation (Royal Decree 1302/86, 26th June on Evaluation of Environmental Impact and Royal Decree 1131/88, 30th September originates from recommendations made by a number of international organisations and from the 1985 "Seveso" directive 85/377/EEC.

It establishes a procedure for assessing certain activities with the potential for pollution. Under the procedure a study to accompany the project must be carried out in certain sites, containing data relating to use of land and other natural resources by industry. This includes an assessment of the possible effects of this activity on the population, wildlife, air and so on, and the measures necessary to reduce or eliminate these effects and suggestions for possible alternatives. The relevant details of whether the project is public or private are
set out in Royal Decree 1131/88 Chapter II, S section 2, articles 7 to 12.

The procedure is set out in section 3 (articles 13 to 22) and includes initiation, information, publicity, environmental impact declaration (which determines the advantages or disadvantages of carrying out the project on environmental effects), and solution of possible discrepancies etc.

Regulations on hazardous activities. These are basically summarised in Decree 2114/61 30th November (which regulates unhealthy, harmful and dangerous activities) and Order of 15th March 1963 (instruction for implementation of the regulation), although there are some rules which are still in force but out dated. It is notable that there is no post-constitutional regulation in this matter.

This regulation defines unhealthy, harmful and dangerous activities and sets out a licensing procedure which is enforced by local authorities, together with penalties on companies who do not abide by the licensing procedure.

Guidelines for Protection of the Atmospheric Environment.
There is a law (Law 38/72 22nd December) which refers to the specific protection of the atmospheric environment developed by means of the Rule by Decree 833/75.

The above law states that the "deterioration of the environment constitutes, without a doubt, one of the greatest problems posed to humanity". It refers to the fact that "the concern about these issues reaches world proportions" and that there is a "need for a General Law for the Defence of the Environment". But this contrasts greatly with the reality of regulations dealing with this matter and with the enormous imbalance and task of modernisation which legislators face.

The Law makes provision for declaring "polluted atmosphere zones" and sets out municipal responsibilities; infringements and sanctions and responsibility for imposing them.

2.3 Legislation on the working environment

Law 8/80 of 10th March of the Workers' Statute, deems "physical wellbeing and an adequate health and policy" (article 4.2.d) a basic workers' right. This basic right compels the state to establish a safety policy and monitor its implementation. Workers have an obligation to self-protection and employers have a duty to protect the health and safety of the worker. The Spanish Judicial Regulation considers the employer as having a duty to protect health and safety at work and employers must:
1) organise work, site/workplace etc. in order that workers' health is guaranteed, taking the necessary measures irrespective of cost.
2) provide the worker with all the necessary measures for safety and protection.
3) inform, instruct and train the worker so as to carry out tasks safely and improve working conditions.
4) supervise the use of protective measures and respect the rules protecting occupational health, and demanding compliance with these rules. Furthermore, these duties are complemented by responsibilities which must be faced in the event of accidents or risks by means of danger money as well as compensation.

This law deems occupational health as an individual problem as well as a collective one, conferring supervision and control on workers' legal representatives within the company. Company committees and workers' representatives are the legal representatives of employees and are able to intervene on matters of occupational health and other issues affecting employees. There are also committees for health and safety in the workplace (in companies of over 100 employees) specifically responsible for occupational health (OGSHT) who are governed by the general health and safety at work legislation (Decree 923/71, 11th March). These committees comprise of both employers and employees. Workers' representatives in these health and safety committees are designated by the company committee. Workers have participation rights and information rights.

Participation rights of a general nature fall into the following sub-groups: a) working hours, working days, shifts (art.41E.T), b) supervision of compliance with occupational regulations (art.64.1.8.b of the E.T), c) supervision of health and safety conditions, d) ability to decide jointly with the employer on situations where exposure to harmful environmental risks should be reduced, e) decision-making in the event of an accident, either by stopping production, as long as the decision is backed by 75% of company workers in companies where there is a continuous production process. This last decision has to be communicated immediately to the management who have to ratify it. f) visit work places to assess conditions and propose preventive measures, g) supervise suitable worker training programmes, h) determine specific risks.

On an individual level, information rights include medical examinations and practical training based on the number of working hours, where the work entails risks to a third party. On a collective level, company committees or staff representatives and health and safety committees have the right to be periodically informed of the statistics on absenteeism, accidents, illness, and their causes, and environmental and prevention studies which are carried out.

2.4 Health Legislation

Traditionally the Health Services Administration (HSA) has ignored its responsibilities in occupational health leaving it completely isolated, despite the fact that the Law of Health Bases 1944 allowed (or at least, was not opposed to) the development of health services responsibility in this issue. It appears that when the law talks of the population's health or of disease control, the HSA assumed that this implied all the population apart from workers and all diseases apart from those which are work related etc.

The announcement of the General Health Law of 1986 generated hopes concerning the possible development of occupational health in the National Health System with a series of explicit functions (art.21.1) and effective coordination between employment and health, explicitly prescribed in article 21.3. Furthermore this law anticipates the decentralisation of the functions of occupational health to Health Area level (art.21.2). It is therefore at this level that the previously mentioned coordination should be carried out, and Town Halls be given responsibilities in the "health monitoring of industry, services, transport and noise" (art.42.3.b). Moreover, the participation of trades unions and employers is defined in the "programming, organisation and monitoring of management related to occupational health, at different territorial levels" (art.22). Another aspect of this law is the obligation of companies to "inform the relevant health authority of substances used in production cycle" (art.21.1f) in order to create risk records. A certain inspectoral role with the power to "enter freely and without previous warning, at any time, into any centre or establishment subject to this law" (art. 31.1.a) is contained in the law.

The transfer of these responsibilities to the Autonomous Communities, since the Constitution allows the "development of basic state legislation" in health issues, gave the Autonomous Governments the ability to act.

Five years after the publication of the General Health Law and with Regional Health Services established in several Autonomous Communities with the exception of Navarra, practically none of the above aspects have been developed, and furthermore, there is a systematic failure to comply with the legal stipulations on the part of Health Service Administration on both state and autonomous level.

2.5 Summary

The first thing which needs to be pointed out is the great extent of standards which exist in this area. The non existence of a general environmental law controlling and assigning duties and responsibilities to different ministries, public organisations, autonomous communities, town halls and social agents, makes intervention in prevention, correction, follow-up and negotiation difficult.

Environmental questions can be included in matters relevant to collective bargaining, either through company committees, or through specific health and safety committees giving the latter more responsibility.

3   COLLECTIVE AGREEMENTS IN HEALTH AND ENVIRONMENTAL ISSUES.

Preservation of the environment, even when its deterioration becomes a serious threat to health, is still not a priority for Spanish employees.

The classic understanding of trades unionism, as defence of workers' rights, despite having undergone important changes, continues to be a prisoner of the wage struggle. This has meant that trades unionism has acquired a monetarist culture which pervades all aspects of the trades union struggle, and from which it is difficult to become separated.

This is not the only difficulty which trades unionism is faced with in its determination to establish a more efficient and competitive unionist model which defends the general interests of workers not just the most pressing ones, and is capable of responding to the latest challenges. Among these, deterioration of the environment is the most important in that it directly affects workers, who are both the cause and the victim of this deterioration in their dual status as workers and citizens. It is in this context that the struggle for health is presented as the most ideal way of raising workers' awareness of the environmental issue.

Of such difficulties three are worthwhile mentioning. One of a cultural nature, another more political and the third of a social kind. The first one is that workers and their unions see themselves ideologically dominated by prevailing medical conceptions about health and disease and are therefore subject to socially established practices. The second refers to difficulties and obstacles encountered when workers try to actively participate within their companies, in an attempt at improving working conditions and the working environment. The third is to do with socio-economic conditions at a time when there is high unemployment, uncertainty at work, along with an ever increasing division in the employment market, all of which condition the capacity of the workers to win demands and the negotiating capacity of their unions.

Employers and the state treat employees' health as an issue to be dealt with by medical and safety specialists rather than a negotiating issue. On the other hand, responsibility for making decisions about working conditions falls exclusively to employers and company directors who have a different understanding in this respect. "At home and at work everyone does as I say" is a phrase often uttered by employers and stipulated in the majority of collective agreements which say "it is the company's responsibilty to determine working conditions and how work is organised".

Legislation setting minimum health and safety standard has been set and enforcement authorities carrying out routine checks, but they do not have effective sanctioning powers. The application of legislation is therefore notable by its absence. This has resulted in participation by workers in preventing health risks being very low. The situation, has improved only slightly from that which existed at the beginning of the 1980s (Perez de los Cobos Orihuel, F; 1991).

The results of a survey carried out by the Occupational Health Board of CC.OO.(Workers Commission) in 1980 are significant:

* Of those companies who are legally bound to have health and safety committees, only 60% comply.

* Of those companies who do possess such a committee, 90% concentrate on purely consultative tasks, without powers to participate.

* 70% of those surveyed consider these committees to be inefficient. One comment was, "all they do is put on an act and allow everything to continue the same as before".

* 80% of committees have no means of monitoring information, and have not received training about the risks to which they are exposed, nor about environmental or health data.

* In most small companies there is no safety supervisor. Where there are, they are chosen by the company itself.

* None of the committees are aware of or have an input into the budgets which the company allocates to health and safety (CáRcoba, A; 1989).

This situation seems to have altered recently according to the results of the National Survey on Working Conditions (1987) carried out by the National Institute for Health and Safety at Work. In this sense the following data are significant:

* Of those companies questioned 36.7% stated that they had a health and safety committee. Of those companies who have a legal obligation to have a committee, the percentage of those who state that they do rises to 73.6% for medium-sized companies and 84% for large companies.

* A third of companies claim that they employ a full or part time health and safety technician. As before, this percentage rises to 58.3% among medium-sized companies and 77% among large companies.

* With reference to the activities of the committees, the survey found that only 21% of companies have a budget specifically allocated to health and safety matters and less than a third of companies have organised training in health and safety for employees (I.N.S.H.T; 1988).

A comparison of these results with the results of a survey by CC.OO. (Workers' Commission) in 1980, show improvements in the organisation of preventive measures, particularly in medium and large companies. For companies as a whole, preventive activity and participation by employees in prevention is still scarce.

Similar considerations and conclusions can be found in the report "Action for Health in the Workplace in Spain" by the European Foundation for the Improvement of Living and Working Conditions (Moncada i Lluis, S; 1991). This is also reflected in the results of collective agreements.

The involvement of employees in prevention activities within companies is achieved by the health and safety committees (OGSHT) where both employers and employees are represented. Other committees, specific health committees, are the result of agreements between employers and employees and their existence ususally shows a strong interest on the part of employees in these issues.

When health and safety is a reason for concern and becomes a clear demand by workers, this is usually translated into agreements on concrete prevention methods which are gathered in collective agreements. Among the prevention measures most often demanded by employees are medical check-ups, preventive actions of a secondary nature, which are also legal requirements.

Collective agreements have legal power, which is why the negotiation of collective agreements can be used by workers to introduce clauses which imply greater respect to those already passed. For example an increase in the rights and responsibilities of workers and their representatives recognised by the law.

If a collective agreement refers only to occupational health and the working environment within the company, it is because the environment outside the company is not negotiable, except in the State Chemical Industries Agreement.

The VII Chemical Industry General Agreement, valid for 1990 and 1991, anticipates in Chapter IX of "Health and Safety at Work", article 55.8, the creation of a mixed committee of health and safety at work and in the environment. This committee has the means to distribute reports on problems relating to health conditions in the workplace and also on repercussions from the chemical industry, in other words, "from on the inside to the outside of the chemical sector". Amongst its priority objectives are monitoring and replacement of dangerous primary materials by less harmful alternatives, follow-up of severe illnesses and changes of technology oriented towards replacing dangerous jobs and staff training activities. This point is significant amongst state agreements as it opens the way for considering the external environment as an issue for collective agreements and constitutes an example of a joint initiative between the industrial actors in this area.

In general, this indicates an advance regarding the usual treatment of occupational environmental questions in collective agreements in Spain. It establishes a periodical register of environmental data with the aim of assuring the maintenance of threshold limits of chemical substances and physical agents; it declares all work which is arduous, poisonous, dangerous or unhealthy as temporary and exceptional, emphasising substitution of safer alternatives it introduces controls for introducing new technology; it formulates workers' information rights and risk supervision and so on.

Situation analysis.

The level of workers' participation in risk prevention in Spain is still very low, despite improvements (Carcoba, A; 1987).

Using the data from table 1 as a basis, several relevant characteristics can be used as indicators of the situation, in order to see developements over the years.

TABLE 1

TREATMENT OF HEALTH AND ENVIRONMENT IN COLLECTIVE AGREEMENTS
Valencia Region and Catalonia 1987. * (1)(2).

	% of Agreements	
	Valencia	Catalonia
Aspects dealt with:		
Knowledge		
Data records	3.1	1.5
Hazard maps	-	2.4
Training		
Occupational health	4.1	-
Right to information	10.2	6.4
Remuneration		
Danger money	24.5	19.6
Explicit rejection		
of danger money	3.1	-
Participation		
Health & Safety Committee	31.6	20.4
Company committee	5.1	-
Specific health committee	8.2	-
Prevention		
No preventative measures	83.7	21.0
Vague general measures	7.1	-
Concrete measures	4.1	10.2
Medical checkups	35.7	22.0
Compliance with legislation		
Legally recognised guidelines	30.6	16.0
Wide legal field	6.1	1.3
Number of agreements analyzed	98.0	382

(1) Own elaboration from work of Unzeta Lopez, M.

(2) Data for 1987 taken from Boix y Ferrando, P 1987.

Characteristics of collective agreements in issues of occupational health and the working environment, taken from the data analysis in table 1:

a) The agreements reflect very little concern on the part of the employees in questions of their subjection to risk.
b) Compensation for dangerous work is widespread only in a few cases are these types of bonuses explicitly rejected.
c) Workers' participation is centred around health and safety, committees to the detriment of company committees which are autonomous institutions with greater negotiating capacities.
d) The majority of agreements in the Valencia Region, over 80%, do not include preventive measures. The situation in Catalonia is likely to be similar.
e) Culture prioritises medicalisation reflected in the fact that 36% and 22% respectively of agreements, propose health surveillance.
f) Collective agreements have very little use in widening the legal field which workers and their representatives have already recognised.

The evolution of these characteristics over time can be established comparing the data referring to the Valencia region and the years 1987 and 1991, tables 1 and 2.

The above comparison, shows that in general things have not improved greatly from 1987 to 1991. It is worth pointing out, however, the small advance concerning knowledge of risk.

There is a greater difference in the situation if the data relating to the State is compared to that for the Valencia region for 1991, table 2. In nation-wide agreements a substantial advance can be seen for the following:
a) Knowledge of risk and especially recognition of rights to training and information.
b) Workers' participation through health and safety committees and the creation of specific health committees.
c) The introduction of preventative measures, 35% of agreements include concrete preventative measures.
d) Negotiation is also being used to widen and improve recognised legal responsibilities.

This indicates that it is the greater negotiating capacity amongst unions on a nation wide basis and in large companies, which gives rise to workers' concern for occupational health and the working environment being reflected in the text of the agreements.

TABLE TWO

TREATMENT OF HEALTH AND ENVIRONMENT IN COLLECTIVE AGREEMENTS. STATE AND VALENCIA REGION 1991.(*)

% of Agreements

	State	Valencia
Aspects dealt with:		
Knowledge		
Data records	5.8	2.8
Hazard maps	11.8	2.8
Training in ocupational Health	29.5	10.6
Right to information	41.0	12.5
Remuneration		
Danger money	41.0	26.0
Participation		
Health & Safety committee	41.0	23.0
Company committee	-	4.8
Specific health committee	17.0	10.6
Prevention		
No preventative measures	64.7	82.7
Vague general measures	11.8	6.7
Concrete measures	35.0	5.7
Medical checkups	47.0	49.0
Compliance with legislation		
Legally recognised questions	29.5	24.0
Wide legal field	35.0	2.8
Number of Agreements	17	104

(*) own work.

Table 3

Employer attitudes to environmental issues.

Ideal positions.

Eco-reticent:
Process of awareness.
* very low - nature = storage/rubbish dump - business predator - rejection and negligence of information

Technology:
* obsolete and polluting technology - conflict between industrial and social ends.

The state and EEC:
* exclusive responsibility for the environment held by the state - protection against impunity due to lack of state control.

Regulations and guidelines.
* premeditated disregard - reticence towards auxliting and payment of levies for dumping - non-fulfilment under the threat of closure.

Environmental industry.
* Unnecessary for company' working as percieved by these businessmen - not considered a genuine concern - domestic operation.

Eco-resigned.

Process of awareness:
* sliight openess - low awareness and acheivement level - pessimistic perception of changes in the improvement of the environment - passive attitude.

Technology.
* confidence in technology innovation - corrective technology - predominance of productive ends - strong depenence on the exterior

The State and EEC.
* victimist attitude - complaints due to lack of facilities - request for more subsidies in order to deal with State demands.

Regulations and guidelines
* begrudging acceptance of environmental costs - fear of rules and regulations - critical of State for not having agreed a moratorium on the fulfilment of European regulations - pessimism about not overstepping the pollution boundaries - fulfilment of regulations puts the financial workings of the company in danger and in many cases it continuation.

Environmental Industry
* resticted use - introduction in Spain in order to reduce foreign dependency - scarce involvement in initiatives.

Eco-receptive

Process of awareness:
* much discussion - high level of information - search for business initiatives - over optermistic perception of the environment issue - active position although the environment is not included in the busness organisation.

Technology.
* technology as a solution for the environment - as a means of strengthening business innovation on the theme of the environment - first steps towards technical research and application guided by foreign companies.

The State and EEC.
* environmental decree between State and companies - recognition of State interest - positive assessment of State contributions (1+D, global willingness, sanctions) - request for positive action.

Regulations and guidelines
* inclusion of environmental costs within the company accounts - need for flexibilty within the regulations. Specific attention to cases - worry about costs - recognition of Spanish industry as opposed to European. Optimism in fulfilment.

Environmental Industry
* use of the services as an ideolgical alibi: make the company image green - environmental reforms disguised as investments initiated by the company - promising sector for new gains.

Eco Active.

Process of awareness:
*internalization of the issue - complex perception of the environmental question - fuision of external and occupational environment - environment another rank of the business system - committed business.

Technology.
* concept of prevention - research and development in the application of innovative and non-contaminating technology.

The State and EEC.
* critical assessment of scarce organisation and framework of a State environmental policy - critique of the lack of infrastructure.

Regulations and guidelines
* high preoccupation with legislation - inclusion of the environment in the company's code of conduct.

Environmental Industry.
* complement to required system of non-polluting technology - habitual use of consultancies, technical studies.

Methodological notes on tables 1 and 2

The report is limited to occupational health and the working environment taken from a sample of collective agreements using previously published studies (Boix; 1987, Unzeta; 1987) the authors own work which refers to available agreements, valid in 1991 and agreed both at State and Valencian Community levels.

Methodological and presentational differences in existing data between the Catalonian study and the remainder have made manipulating the data unavoidable so they are comparable with the others.

The meaning of the concepts which appear in tables 1 and 2 is the following:
Knowledge of risk data (e.g. levels and types of pollution) and data on the effects on workers' health (e.g. accident statistics and sick leave) is imperative for preventive action. Workers can improve their knowledge by negotiation with employers on the use of certain learning tools such as records on noise levels of different jobs, over successive years or on accidents or illness produced in different fields of work.

Hazard maps, this is an Italian methodology for studying and improving work conditions relating to health. They are studies which aim to determine what type of risks exist, where they happen, who they affect etc. and what relation they have with the dangers to health observed. In this way, it is possible to evaluate their relative importance and design a priority plan for preventive intervention.

Risk monetarisation implies the possibility of exchanging health and safety for money, accepting bonuses for dangerous work conditions.

Summary:

Collective agreements on environmental issues are practically non-existent and in health and safety there has been only a slight improvement compared to previous years. This relative incapacity on the part of unions to empower preventative activities in health and environmental issues in companies, using collective agreements has social, political and cultural causes. The above is documented with data taken from surveys conducted on employers and employees and based on the analysis of collective agreements signed in recent years.

4. POLICY STATEMENTS, CAMPAIGNS AND DEMANDS BY UNIONS AND EMPLOYERS

4.1 Introduction

For some years, employers' organisations have been putting together the rudiments of an environmental argument. For the most part, it is still a general, unspecific argument, which has emerged as a response to external pressures and conditions. This has likewise occurred within the main trades unions. In both cases, the argument seems to be maturing and acquiring more complexity and depth since 1990. However whenever the environment is discussed, employers and employees are mutually ignorant of each other. The environment only marginally emerges as an issue in industrial relations.

Employers aim their arguments at the state administration, at other employers and at consumers. It is exceptional for them to consider that workers and their unions could intervene in this type of issue.

Trades' unions in turn, direct their criticisms towards the state administration and, although the blame for the state of the environment is often placed on employers, it is very rare that this type of general grievance leads to concrete initiatives against ecological dangers or pollution generated by one particular company or another. Only very recently have environmental approaches in trades union initiatives appeared concerning collective agreements and tripartite negotiations between the government, employers' organisations and trades unions.

4.2 Employers' organisations.

Although there have been statements, study delegations and so on during the 1980's, employers' concerns on environmental issues are recent. The most significant documents date from 1992. Beforehand the greatest concern was the high cost of adapting Spanish industry. The Spanish Industry Summit, held in January 1990 in Zaragoza, with the participation of the CEOE (Spanish Confederation of Business Organisations) and the CEPYME (Spanish Confederation of Small and Medium-sized Businesses), calculated the investments needed to be more than one billion, three hundred thousand million pesetas. A CEV document (Valencian Business Confederation) from June 1990 maintained that "the new (environmental) legislation compels industry to adapt, the cost of which, on a national level, is estimated at two billion pesetas. The PITMA (Technological and Industrial Environment Plan) from the Industry Ministry, calculates the above investment cost to be 1.2 billion pesetas. According to estimates made public in 1992 by the CEOE, the above investment corresponds to 6.5% of annual earnings of the sector involved. Thirty nine per cent of this amount corresponds to the chemical sector, 31% to the energy sector, 9% to the car industry and the rest to other sectors. Nearly 36% of this investment will have to be made in water treatment, 18% in emission treatment, 15.5% in fuel reduction processes, 11% in dust arrestment, 9% in the development of catalytic converters and the remainder to other activities.

Confirmation of these costs, in company statements, is usually accompanied by a claim for economic help (direct subsidies, financial exemptions, etc.) directed at central and autonomous management. The Zaragoza summit (considered to be highly significant in this area, according to a paper written by the CEOE Environmental Commission's Secretary, Rafael Luengo), demanded the Government provide environmental management with "adequate budgetary resources" in order to "establish every kind of environmental economic support", by means of agreements, either directly with large companies at national level, or with territorial or sectoral business organisations. The CEV document cited that "only with determined participation and financial support from the state, whose responsibilty is unquestionable, can the adjustment be brought about without endangering the viability and very existence of Valencian businesses".

Up until a short time ago, the Government and business organisations had agreed to delay both the adaptation of Spanish environmental legislation to conform with European Community (EC) legislation and their effective and practical application, in the hope that the delay would bring about a marginal competitive advantage for a time. Whilst this delay becomes politically untenable, business organisations increase the pressure on the Government to assume the greater part of investments and costs necessary for the change.

This is fundamentally a question posed in the area of relations between employers and the Government, and not between employers and the unions. However, governmental sources have pointed out the convenience of including this item in three-way negotiations focusing on conciliation.

The PSOE (Spanish Socialist Party) paper known as **Programa 2000 says** "...all the processes described, both the high amount of investment needed to overcome environmental damage and the social importance this issue has in our country, force us to recognise the deterioration of the environment as a state problem. As such, it will only be resolved within the framework of a wide political debate and social accord which involves the whole of Spanish society. In conciliation processes, therefore, it is necessary for this issue to be considered along with its economic implications as a priority problem".

Besides economic help, other claims directed at the State by employers' organisations represented in Zaragoza are the following:
a) although not explicitly, the creation of an Environment Ministry is demanded, by considering necessary "management unity through a medium with an administrative

range similar to that which exists in other EC countries, capable of integrating environmental policy with state, economy, and social policy as is advocated in the EC's Fourth Programme for Environmental Action and the Acta Unica Europea" (Single European Act);
b) the state create infrastructures to clean up rubbish dumps and industrial waste treatment plants, "since competition could become distorted if Spanish industries are subject to the same regulations without the means necessary to comply with them";
c) the state create a pollution control network capable of determining the exact amount of harmful emissions and of establishing a data base for mandatory environmental studies;
d) participation in the initiative established by the emerging "non-polluting industry" sector, defending the existence of companies which include sanitation and industrial waste treatment, along with collaboration between public and private sectors in developing by-product exchanges and re-cycling plants and
e) policy and a legislative framework which empowers the development of the generation of heat from the burning of industrial waste to produce electricity and favours natural gas consumption.

As regards the internal business arena, the guidelines outlined out in the Zaragoza summit were mainly:
- To include environmental management in general management duties (along with production, commercial and financial duties). It is proposed that a specific managerial post in charge of environmental matters be created in large firms with advisory committees for smaller companies.
- To carry out specific environmental audits, oriented towards improving the company's image, increasing its productivity and improving staff training.
- To establish an "innovation triangle" in companies, involving executives responsible for the environment, technical planning and marketing.

The conclusions of the 1990 summit and other employers' organisation papers which have been analysed, the environmental problem appears to be a factor on one hand of state relations, and on the other, although very rarely, as a factor of worker and union relations. This goes some way to explaining the scarce presence of questions concerning the external environment in collective agreements, as has previously been indicated in this report.

So, if in collective agreements, external environmental problems are notable by their absence, in generally termed "green" declarations, exactly the opposite occurs. The bridge between external and internal is seldom crossed, not even when the crossing point is obvious: handling of dangerous substances, noise levels etc. Things like this do not appear to be "environmental problems" in business theory.

4.2.1. Chamber of Commerce Initiatives.

The Chamber of Commerce and Industry (a traditional institution providing services for and co-ordination between employers) has developed a series of initiatives on the theme we are dealing with. They offer consultations to companies by means of expert committees who work for the Chamber. These offices possess a data base on State, Autonomous and Community regulations, they facilitate information about different subsidies and grants, carry out environmental studies, organise meetings and conferences and so on.

One of the most relevant activities has been the development of the so-called Plan Cameral (Chamber Plan) covering: educational policy, training, help and advice to the PYME, state-chamber of commerce working groups, data bank, by-product exchange and dissemination and adoption of initial guidelines on environment for industry.

In educational policy two activity channels are proposed. On the danger of certain products and production techniques, and close collaboration with the State to negotiate of financial support and an adjustment period to new needs in environmental matters.

The training stage is mainly centred on disseminating activities aimed at training environmental experts, establishing alternative technology for each industrial sector and encouraging the inclusion of environmental specialisation as a University discipline. It is expected that this training programme will be carried out in collaboration with the state and the universities.

As far as advice and help for small and medium-sized businesses (PYME) is concerned two aspects can be distinguished, one of a legislative nature including offering advice to companies on matters of environmental regulations, and another financial aspect informing company executives on matters of subsidies and credit that allow development of new investments.

State collaboration can be found specifically in the Plan Cameral, which proposes a working group consisting of two representatives designated by the high chamber council and others, and two representatives from the Public Works and Town Planning Ministry. The working group would be responsible for developing regulations, finding out the costs of new investments into "clean" technology, creating funds to deal with EC legislation and developing a follow-up to the adaptation of different industrial sectors to EC legislation.

The previously cited stages will be preceded by the establishment of a data bank on financial and legislative matters, allowing the private sector to find the level of investment in environmental control, the cost involved and information on new technology. Such information will be strictly confidential, any intervention by the state will be rejected given the distrust that could arise in the business sector.

The creation of a by-products exchange becomes a necessity when faced with the application of legislation on dangerous and toxic waste, and the implementation of a National Plan for treatment of industrial waste. The by-product exchange allows contact to be made between buying and selling companies, in such a way that the latter would avoid paying withdrawal and treatment tax, and the former would obtain prime materials at a reduced cost. The business sector reiterates the importance of state grants which would facilitate the by-product exchange's creation and the development of the previously mentioned stages of the Plan Cameral.

4.2.2. Policy versus practice

Public image in company policy, in fields such as publicity and dissemination of information, is becoming important with regard to the environment. But there is sometimes a contradiction between public image and company practice. For example the Spanish Association of Aerosol Manufacturers, because of the strong presence in the mass media of information about ozone depletion and the ratification by the Spanish Government of the Montreal Convention, edited and distributed material justifying the role of this industry and presented it as being highly aware of its responsibilities. In 1990 a widely distributed leaflet assured consumers and the public that "more than 90% of aerosols manufactured in Spain do not contain CFC'S", and that the industry was 10 months ahead of schedule in complying with international committments.

This publicity campaign tried to combine an image of this industry as responsible and concerned with the need to stop damage to the ozone layer, with the insistence of multiple use of aerosols in order to increase the consumption of their product. No mention was made of the negative ecological effects which have been detected in some propellents used as a substitute for CFC'S.

This campaign is an example of the change of tactics in certain sectors. No longer do they deny or ridicule the ecologists arguments, instead they attempt to present an ecologically-sound image of the company or sector in question. Although the campaign leaflets insisted the breakdown of the ozone by chlorine in CFC'S "has not been totally caused in the laboratory process", the emphasis was placed on publicising the rapid fulfilment of commitments demanded by the Montreal Convention.

4.3 Union organisations.

Union forces have opened the way for formulations on environmental matters, in part owing to the extent of public opinion on the green movement. From the 1940's until 1975 (death of Franco) official trades unions were not by definition organisations which made demands in areas such as improvements in working conditions. For their part, secret trades union organisations concentrated on the fight for freedom, and, above all, for workers' rights, from the point of view of wages and maintaining jobs.

It was at the end of the 1970's, once unions had been made legal, that they began to take on claims from other social movements including pacifists, ecologists and feminists. With regard to the environment, these early years saw them opposed to the lack of effective legislation on pollution, nuclear power stations and so on. There have also been numerous initiatives on working conditions which have environmental implications for example on noise and dangerous substances.

The VII FEMCA-UGT Congress (1980) passed a resolution on "Occupational Health" in which the federal executive was ordered to develop a project on the controlled use of asbestos. In 1981 this union published a text entitled "asbestos kills", and were threated with libel action by the employers. There has also been union involvement in citizen mobilisations provoked by the high pollution level caused by industries in districts inhabited by workers. The most well known was in Erandio (Basque country), where in 1969 there was a powerful conflict caused by gas leaks from factories in that zone, which lead to strikes).

In recent years there has been a tendency among the trades unions to address environmental problems are less directly connected to living or working conditions. For example, references are made to deforestation and consequent geological erosion, pollution of the Mediterranean and there has also been union involvement in campaigns against nuclear power promoted by the green movement.

At policy level at least, there are tendencies to prioritise the environmentalist perspective. In the resolutions made at the International Environment Conference held in Barcelona in February 1988, organised by the inter-regional Pyrenees-Mediterranean Trades Union Council, with the UGT represented by Catalan delegates, it is maintained that workers should reject dangerous work and should not accept technology which is too noisy, involves high temperatures or exposure to toxic substances.

The positions of the UGT (General Workers' Union, a traditionally socialist trades' union), and CC.OO. (Workers Commission, a traditionally communist trades' union), the two largest union federations are examined below. (The less influential organizations, such as the CGT anarchosyndicalist tradition has for many years been open to receiving ecological approaches).

4.3.1. Comisiones Obreras (Workers' Commissions)

This union has been holding meetings between trades unionists and ecologists, since 1978, with the purpose of debating and finding out their respective points of view. Even before being legalised in the 1970's some factions linked to the union participated in people's protests on pollution or poor environmental conditions in workers' districts in numerous cities. Sporadically, there have been initiatives to denounce poor safety or hygiene conditions in the workplace (Quash-Tierras in Almeria, FYESA, in construction and nuclear power plants), to denounce pollution caused by industry (Bazán de San Fernando, bahía de Algeciras, on hospital waste, incineration for example), often linking workers' health with the impact of the external environment (paper factories in Miranda de Ebro and Ferronor).

At the National Conferences of 1983 and 1987 motions relating to occupational health and work conditions were debated. In 1983 reinforcement of occupational health committees was demanded, attributing them with the capacity to carry out research into the working environment. In 1987, the deaths of 26,299 employees in accidents at work in the previous ten years were denounced and numerous measures for improving work conditions were demanded.

In more restricted territorial areas, CC.OO. conferences have occasionally debated ecological problems. The 1983 CC.OO. Congress in the Valencia Region, for example, presented two basic lines of action in this field opposition to nuclear power, demanding less polluting and cheaper power; and the demand that unions be able to take part in areas such as town planning and motorway design.

The 1986 Congress in the same region introduced an analysis of economic, agricultural and service costs. It proposed organising courses and study conferences to raise workers awareness of ecological issues and dispel possible fears about the introduction of "clean" technology, with the aim of convincing trades union delegates that this would not mean job losses. It advocated the creation of a study commission made up of experts in environmental matters and denounced the serious deforestation of land and the increase in dumping of waste in the Mediterranean.

Union spokespersons place particular importance on environmental initiatives included in the ISP (Union Progress Initiative), a platform established jointly with UGT. This document refers to energy problems and the management of environmental resources. In the first field, the relative inefficiency of Spanish industry is described as a competitive disadvantage, and energy planning which favours energy saving, "clean sources" and strengthens "renewable energy" (the production of electricity from industrial waste fuel) is demanded. The report defends the indefinite moratorium affecting the construction and operation of new nuclear power stations and recommends a drive towards renewable energy sources and research into "clean" coal combustion processes. And it advocates compatibility between reducing energy consumption and economic growth. In the second field, the principle of sustainable development and planning for use of natural resources is demanded and the trend of postponing or reducing levels of compliance with pollution controls demanded by the EC is condemned. It demands the strengthening of the rail network, a Natural Resource Law which determines and widens the range of ecological offences, an inventory of natural resources and the conversion of highly polluting industries without job losses.

However the CCOO only formulated programmes on environmental issues at the highest level at the end of 1991.

In the above Congress, the general secretary was opposed to the union accepting the creation of highly polluting companies in exchange for the corresponding jobs. He favoured "sustainable development" and demanded the conversion of occupational health committees into environment and health committees.

The Congress also passed two resolutions, on "energy policy" and "union action and the environment". The first was in favour of energy saving and the use of renewable energy. The Congress was also in favour of the closure of all Spanish nuclear power stations over the next ten years, "The problem can no longer be considered by means of more moratorium. Faced with the confirmation of pollution caused by nuclear power stations, the enormous volume of radioactive waste they generate, their applications in the military field, their high risk of accidents .... the only response is renunciation of such a source of energy" (CC.OO., 1991:3). The second resolution demanded the right for unions to participate in and be consulted at all decision-making levels, concerning health and the environment and proposes the creation of a network of union environmental representatives in small and medium sized businesses.

The Congress also voted for the creation of a department of ecology and environment within the union. This department has taken various initiatives and has promoted contact between the union and a wide range of ecological groups.

In the criteria for collective agreements for 1993, the CCOO suggests the possibility of introducing clauses committing companies to facilitate information and accept union participation in environmental matters. It proposes, also, the carrying out of eco-audits, adaptation plans to conform with EC legislation, energy saving, water purification and waste reduction, and so on all with the participation of the union.

4.3.2. General Workers' Union (UGT)

The UGT has established a framework for its environmental policy resolutions made at its XXXV Confederal Congress (1990). In general, ecological concern is introduced without perceiving a fundamental conflict with commitments relating to economic growth and employment. The "ecological defence of the environment" is defined as one more element of "new industrialisation" which guards against the "false contradiction between ecology and economy". The resolutions maintain that the "the polluter, pays principle" is insufficient and priority should be given to pollution prevention; that environmental training of union members is necessary; that an environmental section in the leadership of the union should be created. The resolution also announced the notice of a future conference on trades unionism and conservationism and denounced governments who, regardless of their ideology, have allowed pollution, toxic chemical products, indiscriminate felling of forests, development of nuclear energy, and so on. These denunciations are linked to the lack of preventive measures to protect workers who have to handle polluting products.

The union sees protection of the environment as an opportunity for job creation it demands a General Environment Law and the creation of a State Council to advise the government on issues such as the non-production of toxic waste and the replacement of nuclear energy with alternative forms. The resolutions of the XXXV Congress include an informative annex on the industrial origin of pollutants such as SO_2 and NO_x, ammonia, surface ozone, heavy metals, dust, and CO_2 and their effects on eco-systems and human health.

The UGT is in favour of more industrial expansion and more economic growth, with the proviso that companies apply necessary measures to curb environmental deterioration. The tone of conciliation between both elements is very high. It has even been said that industry ought to maintain present production without "any pollution or wastefulness".

The occupational health resolutions at this same Congress attempt in some way, to connect the working environment with the external environment. They demand the right of workers to self-protection, information and participation in everything relating to the definition of working conditions. They request a General Occupational Health and Working Conditions Law to implement the EC Framework Directive 89/391 and International Labour Organisation (ILO) Agreement 155. The trend of subcontracting dangerous jobs is denounced and it was pointed out that industrial pollution affects areas and districts where employees live more intensely. It sets out that all the aforementioned should be translated into collective bargaining and into action amongst companies by means of occupational health committees (preferably in large companies) and of territorial health and safety representatives (centred more on smaller companies). These committees should replace existing ones and monitor pollution control and so on.

Union tasks with regard to the environment are summarised as: acting as a social and political sensitisation agent together with other similar organisations to reorient economic order and contributing to society as a whole and to organisations defending the environment.

The aforementioned resolutions have begun to be translated into practice. The UGT's instructions for collective bargaining in 1991, recommend that agreements be included relating to information and consultation on all environmental issues which companies are involved in; participation in decision making on changes in production processes and investments, monitoring of waste; knowledge of compliance with environmental legislation on the part of the company and all levels of company policy to have an environmental aspect.

The UGT has organised varied debates and has designed a course on "health and the environment" for its staff training school.

The issue has also emerged in sectoral and regional programmes. Thus, in 1990 in Extremadura the UGT announced that it was against nuclear plants on regional territory, and in favour of conservation of indigenous forests. That same year, a state federation congress of chemicals and energy connected occupational health matters with the external environment. The UPA (Small Agricultural Union) linked to UGT, is in favour of strengthening biological agriculture and in general, cultivation techniques which reduce pollution.

At these more specific levels, however, tensions begin to show. The Extremadura congress criticised those ecologists who claim to maintain the region as a "bird airport". Union representatives from the chemical industry contrasted the growing social concern for the environment with the scarce concern for workers' health: "regulations relating to protecting the environment are much stricter than those pertaining to the defence of citizens' and workers' health... it seems strange and incredible that society, which goes deeply into improving the environment, does not assume the defence of workers' health with the same emphasis".

4.3.3 The Environment in the Priority Union Proposal (PSP)

In the preparation of the 14 December 1988 general strike, and since then, there has been a powerful tendency to unite in action between CCOO and UGT. The 1989 Priority Union Proposal (PSP) outlines the joint position of both unions in view of negotiation with the government and the employers.

Although the environment occupies a secondary place in the PSP, it is the first time that a union platform concerning immediate practice and planned at such a high level contains relevant environmental criteria.

The PSP advocates public subsidies for investment and research, so that environmental impact can be considered a priority in relation to economic-monetary interests. It demands that workers' rights should be recognised so that they be consulted before decisions are taken concerning investments or changes to the production system. The right to information on the degree of compliance with environmental legislation is also demanded. The PSP defends the application of the principle "the polluter pays" along with the immediate stoppage of any activity which generates a high level of pollution. Finally, planning is defined for growth, involving workers and unions, inspired by the prevention (not repare) of ecological damage.

4.4 Summary

Until recently, employers' organisations trades unions had paid very little attention to environmental problems. From 1990 onwards, the situation has begun to change and there are more and more programmes and statements, and the arguments are increasingly complex.

The main point of interest for employers is the adaptation of Spanish industry to the EC environmental legislation. They demand help from the Government to pay for this adaptation, improvements in the coordination of the various environmental authorities. They propose the integration of the environment into management functions of companies.

Trades unions have denounced the dangers of the nuclear power industry, the risk of desertification in Spain and the pollution of the Mediterranean for example, tending to adopt the positions of the environmental movement. But only on very few occasions have they included environmental grievances in their negotiations with companies or or the government, although this is changing slightly.

Employers consider environmental management to be their exclusive responsibility and that unions should not have decision power in this field. Unions, for their part, demand rights in this area, despite the fact that efforts to achieve this in practice are very scarce.

5. ATTITUDES OF INDUSTRIAL RELATIONS ACTORS TOWARDS THE ENVIRONMENT

5.1 Introduction and methodological clarifications

Analysing the attitudes of industrial relation actors in Spain entails numerous difficulties owing to the limited information sources. On one hand, reports on attitudes and opinion carried out by business or trades union organisations are scarce and of a fairly irregular nature and often give a 'snap shot' picture rather than reflecting long term practice. This is the case with a number of surveys carried out at the request of employers' organisations. These studies survey very specific aspects and in many cases do not allow sectoral or regional generalisation to be made. There are recent studies on attitudes to EC legislation or on the inclusion into the company of staff specialised in environmental matters but these are not published and their circulation is often restricted.

Secondly, surveys on attitudes and opinion which are carried out by the general Environment Secretariat, resolve to a certain extent the problem of irregularity, as since 1986 three large surveys have been carried out (1986,1988 and 1990, the latter completed in February 1991).[3]

In this way, the information available allows comparison as the questionnaire used is virtually the same on all three occasions. However, these surveys present some obstacles firstly, on dealing with a survey aimed at the entire population, difficulties arise when it comes to differentiating between employers and unions. In the case of employers, it is possible to separate their attitudes from the rest of the sample thanks to the responses indicating profession of the person surveyed (although this is not possible in the most recent study). Also the design of the questionnaire, conceived for the population as a whole, is not ideal for specific themes, and there are problems with the representativeness for employers within the total sample.

Other document sources have therefore been used including: publications by claimants, declarations, proposals and programmes, conversations and interviews, press archives. It has therefore been possible to look more in depth at the attitudes of Spanish employers and unions and evaluate how they are applied in practice. Press archives have been very useful[4] as they have allowed an exploration of the conduct of industrial relations actors, allowing them to speak for themselves and comparing the level of declaration with the level of practice.

---

[3] See IDES: Estudio sociologico sobre medio ambiente en Espana. Madrid. 1986. EMOPUBLICA: Actitudes de llos espanoles hacia el medio ambiente. Madrid 1988. RABIDA CONSULTORES: Opinion publica y Medio Ambiente. Madrid. 1991

[4] Collaboration of those in charge of Environment Secretariat documents service, National Hygiene and Safety Institute and 1st May Foundation (CC.00) facilitated our work very much.

5.2 Attitudes of employers towards the environment.

Documents from employers organisations aimed at spreading environmental good practice and rules of conduct amongst employers and promoting individual company policy have been reviewed. And information has been used which appeared in the Spanish press, mainly from March 1991.

These two document sources, one of a more "formal" nature and manifest in the business environment, and the other closer to actual events provide a view of employers' attitudes towards the environment whose essential feature is duality, or even, paradox.

Employer policy tends to underline solidarity between producers and consumers concerning protection of the environment, and the harmony between development and the environment by means of innovative technology. The paradox is created between values which make up this argument and the production process itself, which is often revealed by the media as environmentally damaging. The protection of the environment is used by companies, particularly in highly polluting industries such as chemicals and energy as something "that sells" and tends to create a modern and competitive image.

The common feature which characterises the employers' position on environmental issues, is "the insistent inclusion of this variable in their management strategies". The change of focus moves away from defensive stances maintained previously in order to include new elements of initiative and coordination.

Industrial growth is not questioned, but it has to be linked to respect for and commitment towards the environment, even more so when social demand is growing in this field. The question that employers pose is how to manage in a world concerned about ecology and this leads them to prepare themselves for a clash against the destruction of the planet, to look closely into the costs of pollution and to extol traditional values tinted green.

The following references illustrate this remarkable change, and are examples of employers connected with particulary polluting sectors (chemical, power, car industry):
> "It is possible to state that, socially speaking, we are submerged in an (ecological era), on which no society, country or community can turn their back" ("Empresa y Medioambiente", Bulletin n° 52; Page 44. President of Administrative Council SEAT).

> "The attitudes of employers as regards the economic role of the environment has evolved in a rapid and important way. The old fashioned idea of considering conservation as an unproductive burden has come to be a (socially responsible committment) which implies efficient management of programmes and resources" ("Empresa y Medioambiente", Bulletin n°

52; Page 140, Administrative Commission REPSOL QUIMICA).

Along with this, it is considered that industry is an essential part of the protection policy of the environment, the vanguard of progress:

"I think that the industrialist of our time should be aware of the degree of deterioration which our planet is undergoing and although there are those of us who know that industry is just one more cause, and not the main one, this should not be anything other than a stimulus for being the vanguard of environmental protection. (Large industry can become the locomotive that, with its example, can pull the train in which the state will undoubtedly have to place the necessary energy)" ("Empresa y Medioambiente", Bulletin n° 52; Page 114, President of CEPSA).

However, the above mentioned stances often clash with practice. There are numerous examples of recurring infringements of standards. Illegal dumping is so widespread that compliance with the regulations is an exception. Everyday practice does not comply with the law, and the pollution continues even after companies have been reported, sanctioned and fined.

There is a trend towards proposals for corrective measures, which instead of requiring modifications to the production process, displace pollution to another location. For example a number of paper factories base their waste strategy on disposing of industrial waste out at sea.

The case of dumping is the one which most often demonstrates anonymous conduct, when it is difficult to determine the polluting substance after it has been dumped. The case of the Albufera de Valencia can be cited here, which suffered an unknown spillage in April 1991 causing the death of thousands of fish ("Ya" 24.4.91), and the case of the Gernika estuary, which had similar consequences, on 2nd April 1991 ("Correo Español" 3.4.91). Two toxic spillages took place in the same estuary in June 1991, but it has proved impossible to clearly identify the substance which caused the pollution ("Deia" 12.6.91).

There is a problem of lack of reporting to authorities. This practice is exemplified by non-declaration of dumping within the time limit set by the authorities. The lack of reporting occurs where normal regulations are involved and in accident or emergency situations.

The annual production of toxic waste by Spanish industry substantially exceeds the capacity it is able to sustain. (According to MOPU, toxic and harmful waste production in 1989 was 1.8 million tonnes. On the other hand, by 1995, capacity for sustaining waste will be only 200.000 to 400.000 tonnes, 20% of the present level). Despite this some companies still import toxic wast from abroad.

There have been a number of cases where companies have shown no cooperation towards the state in developing campaigns for protecting the environment and have avoided their responsibilities.

For example, the battery collection campaign called "botón", organised by the General Environment Secretariat this year following the "polluter pays principle", needed manufacturers' cooperation in collecting batteries but met with a very negative response despite the low cost of this initiative ("El País" 18.3.91).

These contradictions between stated policy and practice are accompanied by a profound defensive attitude. For example employers deny that processes cause environmental damage or argue that damage is minimal.

This attitude is usually supported by scientific arguments to refuse investigation which offers contrary conclusions. Sometimes the argument is presented in comparison with other damage socially considered "normal". Some employers minimise the damage that their activity is causing and compare it to car pollution or central heating systems.

When recognition of an infringement becomes impossible to avoid, some companies have justified the infringement in accordance with three types of argument:
    A. - Failure to report to authorities.
    B. - Lack of sites to deposit waste.
    C. - Lack of economic resources for decontamination (accompanied by a request for State financing).

Many industrialists consider that financing the changes that Spanish industry has to go through should be a priority responsibility of the State.

Companies have also attempted to evade responsibilities for environmental damage. Employers have argued that manufacturers and consumers are equally responsible for pollution (although clearly employers have the power to change production processes). In this vein the claims of José María Cuevas, President of the Spanish Confederation of Business Organisations (CEOE), are of great interest. "It concerns the whole of Spanish society, **including employers**". "Expansion" (22.12.90) The claims of the most important representative of Spanish employers do not show much determination to take on this responsibility.

Employers are managing to evade compliance with environmental legislation, by means of threats of jobs losses if the company were to close down. Fear of job losses caused by closure of a polluting site, appears to have an effect on the Governments' caution when it comes to creating drastic measures for companies.

The official employers' argument proposes the signing of an environmental pact, as a political measure to undertake the reconversion necessary to environmentally clean up industry. This says "Our country is still not organised well enough administratively to promote and coordinate environmental policies and plans. However, major infrastructure deficiencies and leave Spanish industry at a disadvantage when it comes to complying with more and more demanding community regulations. For this reason, the promotion of "environmental pacts" becomes more and more necessary, so that solutions can be based on shared responsibilities between legislators, regional and municipal authorities, industry and, finally, consumers" ("Empresa y Medioambiente, Bulletin n° 52; Page 143 President of Administrative Council REPSOL QUIMICA).

The pact is understood as a way of nationalising responsibilities with other agents, of acquiring leadership, and of establishing subjects for negotiation (unions and "green" organisations are excluded from this type of proposal). Likewise, the pact could be used as a way of channeling different types of aid requested of the State.

However the majority of agreements achieved do not exceed a nominal level of committment and are without agreements or concrete guarantees. For example in Tarragona, where the local authority, the Chemical Employers Association, Asociación Química de Tarragona and fifteen companies from the chemical sector located in the area, signed a document of "responsible conduct" in February 1991, in which companies committed themselves to reduce pollution to the minimum, to improve security measures, and to collaborate with the authorities in informing the public about the potential effects of the chemical industry on the environment. ("Diari de Tarragona" 8.2.91). The Mayor of Tarragona, "recognised that the document could have been much more committing for industry". Together with the non binding nature of the commitments, the parties who signed this document are companies and local authorities, with the exclusion of other social bodies such as neighbourhood associations, unions, and green organisations.

Technology is a key concept in the environmental argument among employers, as it is only by technological innovation that a link between development and the environment can be created. The same argument marks the path of a traditional technological concept, defined as a means of dominating and transforming nature towards a new concept, where technology is conceived as a means of balancing production with environment through innovation in manufacturing, material and re-cycling processes. The traditional concept of technology considered deterioration as a necessary cost, inherent in the industrialisation and economic development process.

But the new image of technology which employers defend is about combining economics and the environment, "An alternative use of technology will serve to repair forests and rivers, to condition marshland for breeding aquatic birds, to adapt new and existing public works not to disturb animal development, to repair indigenous woodland, to make wildlife return to our rivers, reduce wildlife mortality from plagues or epidemics; to sum up, to restore the planet to equilibrium. ("Empresa y Medioambiente, Bulletin n° 52; Page 207. Administrative Council ERCROS).

Employers are optimistic about technology resolving present environmental problems in the future, offering a technical solution to environmental problems. However, despite the existence of alternatives to polluting production systems, in many cases technological solutions have not been developed for dismantling sites constructed in the past. This is the case of a uranium factory in Andújar, whose closure plan was opposed by the local community and ecologist groups, as they considered shutting down the factory could increase radioactivity in the zone. The company ENRESA recognised this in its dismantling plan ("El Independiente" 11.4.91).

The same has happened with the nuclear plant Vandellós I, where the high cost of dismantling (around one hundred thousand million pesetas), combines with the fact that all the structural materials of the plant have undergone a process of radiation, which signifies the generation of enourmous quantities of radiactive waste. In Spain there is no specific regulation for dismantling. ("ABC" 3.4.91).

So although technical solutions can offer better environmental management, the alternatives for high polluting processes are still limited. Technology, alone, does not deal with the great quantity of toxic and harmful waste which has already been produced and continues to be produced.

Also, the so called "new non-polluting industry" does not consist of replacement of polluting production processes with non-polluting ones, but rather it is a sector which concentrates on producing purifying and recycling systems, far from eradicating the presence of polluting industry, bases its cost-effectiveness on this industry's existence. Thus, it is found that the majority of companies who create subsidiaries dedicated to environmental engineering cause pollution themselves. And even where facilities such as waste disposal are available, companies may still illegally dispose of waste. "In 1983, there were three treatment plants, but within a few years two of them closed down due to lack of cost-effectiveness" ("El Independiente" 26.2.91).

On the other hand, despite the expectations expressed by employers with respect to the opportunities of technology in environmental protection, there is no sign of a transformation within the organisational structures of companies. A survey carried out at the end of last year, to find out about specialist staff assigned to this area within the company, shows that only 29% of Spanish companies entrusted this task to high executive posts, and 13% placed it in hands of personnel departments. ("La Gaceta" 6.10.1990).

Frequently, companies consider that the authorities are responsible for a major part of the costs of pollution reduction or relocation projects. In the case of PEÑARROYA (a chemical company), which had to invest 4000 million pesetas in the construction of a new sulphuric acid plant, the directors believe a subsidy of 50% is necessary. Management and unions are of the opinion that "without subsidies, the multinational group Metal Europ, which controls PEÑARROYA, will not only not build the new plant, but will not even keep the present factory structure in Cartagena". ("Cinco Dias" 17.1.91).

In the removal of factories which are situated close to the urban centre and store dangerous waste, businesses often ask the authorities for indemnities. For example, QUIMICA IBERICA (another chemicals company situated close to a school in San Fernando de Henares) directors refused to dismantle the factory without first negotiating an adequate indemnity. ("El País").

The transformations which Spanish industry will have to undergo in order to adapt itself to environmental legislation are varied and numerous, concerning employers. The Energy and Industry Ministry estimates the cost of adapting Spanish production at 1.2 billion pesetas. The paper and chemical sectors would absorb 39% of the total investment and the power sector 31%, car and construction would also take up 9% and 6% respectively.

Employers request a flexible attitude by the State when the time comes to implement the regulations; "It is necessary to underline this serious problem which could arise from the establishment of generalised demands and immediate implementation. It is foreseeable that if the same demands are required by each company, problems of adaptation will be numerous both in costs and in possibilities of tackling them, which in turn could cause serious harm if excessively rigid action is favoured. [1] (Page 181 Presidente del Círculo de Empresarios Vascos) (Basque Employers' Organisation).

Companies also criticise the state for failing to provide suitable sites for industry, "The country must offer similar infrastructures to those of other community countries in as far as rubbish tip sanitation and industrial waste treatment plants, since, if not, Spanish companies will be disadvantaged if they are required to comply with the same regulations but with less means of doing so" ("Empresa y Medioambiente", Bulletin n° 52; Page 141, President of REPSOL QUIMICA).

In the case of smaller companies, like QUIMICA IBERICA, criticism of the State for absence of toxic waste incineration plants, is cited as the reason which makes compliance with regulations impossible, because the low volume of waste generated makes its possible treatment expensive.

As regards employers' attitudes towards EC regulations, the newspaper "Expansión" (15.5.91), cites a survey carried out in petrochemical, mining, plastics, car and iron and steel

sectors; all of which are greatly affected by the implementation of legislation. According to this newspaper, the survey contributed the following data:
- 29% of industry feels that new legislation will have negative repercussions;
- 24% think that it will affect them positively;
- 35% say they are unaware of environmental legislation and
- 76% state that they have plans to improve their environmental situation. Of these, 54% within a period of five years, 33% in one or two years, and 13% within less than a year.

In summary, with regard to the technolgical concept, there was optimism, perceiving infrastructure reform for less polluting production as a challenge. With regard to legislation the attitude is one of fear and pessimism. Many reservations appear when it comes to applying pragmatic plans to non-polluting technology.

The investment Spanish industry needs to make in environmental conversion to prepare for EC environmental legislation in the next two years, is considerable (1.2 billion pesetas). This has given rise to the industrial world seeing in this reform an open field for business.

However, very few Spanish firms offer services in this field. According to the Minister for Industry and Energy in the magazine "Mercado" (2.1.91): "Spanish industry is insufficiently supported". At present there are 300 medium-sized companies who could be the embryo of this type of industry. According to this magazine, the principle characteristics of this industry are scarce specialisation in the industry in team working, scarce specialisation by engineering companies, who could channel an important part of their business into environmental activities; and the absence of companies who could contribute to systems, to prevent or treat waste products. According to "Cinco Dias" (17.3.91), foreign companies have the most opportunity to enter the Spanish market, due to the wide range of environmental services on offer.

However, there have recently been Spanish investment projects: the opening of the company "Tecnoambiente", by the EULEN, IBERDUERO, Diputación de León and the Spanish water company group, whose aim will be to introduce environmental activities related to the water cycle. ("Expansión" 21.2.91); or the new subsidiary of the chemical group ERCROS dedicated to the promotion of environmental industries and services (Prisma) ("Expansión" 20.2 91). Regarding waste treatment, in May 1990, "Emgrisa" was created, a public company centred on treatment of industrial waste, a State initiative set up in view of the lack of private initiatives ("Cinco Dias" 28.1.91).

Despite these examples, there is no business development suitable to confront the needs of industry in this area. Moreover a trend can be perceived on the part of companies belonging to the most polluting sectors, towards the creation of subsidiaries dedicated to the business of "eco-industry". This does not end polluting activities, but merely provides more benefits for the parent company both in economic and symbolic terms.

The press has reported that employers are making investments in "clean up" measures. But where these are dictated by compliance with regulations or sanctions, it does not mean that they are part of a preventative environmental concept which guides the planning of production. For example "Río Tinto Mining invests four thousand million pesetas in the environment" ("Expansión" 4.3.91), refers to with investments in work being done on a slag treatment plant, and repairs aiming to reduce the emission of sulphuric acid. These are internal corrective measures within the Huelva Corrective Plan carried out by the Junta de Andalucía's environment agency and thirteen large and medium-sized companies. This plan involves the investment of ten thousand million pesetas, 95% of which corresponds to internal corrective measures by companies, standing out amongst these Rio Tinto Minera and the Empresa Nacional de Celulosas (ENCE).

"The Basque investment into the environment will be 95.000 million pesetas" ("Expansión" 11.4.91) refers to investments for adapting to European environmental legislation, according to a Confederación Empresarial Vasca plan. According to this plan, the main expenditure will take place in paper, chemical and steel work sectors.

"Eighteen companies invest 19.000 million pesetas to end pollution" ("Diario 16" 11.1.91) refers to a case where the majority of investments are those undertaken by the companies FESA, PEÑARROYA, REPSOL PETROLEO and PORTMAN GOLF, all situated in Cartagena, one of the industrial pollution blackspots in Spain, and all associated with environmental infringements to such an extent that two of them have been repeatedly closed this year. The companies will count on anticipated subsidies from the Industrial Environmental Plan, of the Industry and Energy Ministry, and will be carried out over the next three years. So the main motive of Spanish industry's environmental investments is compliance with, fundamental EC legislation.

Employers frequently have the difficulty of reconciling their economic interests with the environment issue, and this leads to contradictory actions.

Table 3 is a summary of the types of employer attitudes to environmental issues, based on "ideal" lines of action which represent conceptual models and combine business practices.

5.3 Union attitudes towards environmental issues

The wave of complaints about working conditions in Europe in the 1970's reached Spain a littlelate. However, the efforts of unions to produce successes in health and improvements in working conditions has rapidly increased the level of discussion about this issue. The concept "working environment" and its linkage to a complex argument with external environmental problems, is still being debated.

The adoption of an environmental language, does not imply the internalisation of its meaning or its verification in practice. In the case of unions, the extent of concern about the ecological issue is growing and the strengthening of this "new solidarity" may strengthen traditional union action, and initiate links between outside and inside of factories.

Although pacts have been established between the two largest union federations CC.OO and UGT,[5] there remain divergent stances can be observed among leaders of each organisation. The differences in opinion between these and the activists or group of workers are shown, fundamentally in cases where the latter see their jobs threatened by pollution and closure.

There are two trends in union positions "ecological" and "pragmatic" trade unionism. These demonstrate the ideological situation and union action and depict a tension within trades unions which may give way to important innovations in attitudes.

The first trend is represented by minority sectors, even by personal actions by those in charge of areas relating to environmental issues. However, the growing importance of ecological content may make the second tendency increasingly open to these propositions. The second focus (pragmatic), is predominant and maintains a point of view which is more connected to the working environment and unwilling to abandon a stage which has produced important victories and which is perceived as particularly adapted to union action.

"Ecological" trade unionism develops a more innovative argument within the trades unions, integrating issues of health and work within a global concept and removed from the environment. Only after criticism of conditions which worsen the state of the surroundings and natural resources, will it be possible to pose an integral conception of health prevention in workers. This approach suggests the incorporation of political ecology into trade union action, with the consequent tensions produced, specifically criticism of productivism and of the capitalist economic growth model.

---

[5] In recent years we have witnessed, if not complete unity between UGT and CC.OO, joint action in many aspects. This collaboration, although it does not dispel particular visions of each one, allows lines shared by both and declarations made to be used in this report as principal documents for the study of attitudes. This willnot lead to a specific study of each organisation, but to proposals which originate from their present joint activity and the tensions they both share.

"The present social system of capitalist production is irreversibly destroying the health and natural basis of human activity. Undoubtedly, the ecological crisis forms part of one of the dominating systems of the present crisis in the capitalist system. The impossibility, which has been scientifically demonstrated, of spreading the capitalist social and economic system to the planet as a whole implies the confirmation of the need for a profound change in the economic industrial system and a simultaneous transformation of the social models which support it". (Eduardo Gutierrez, Area de Medio Ambiente. Fundación 1 de Mayo).

In this sense, the starting of a system which takes account of environmental issues, with resources for cleaning-up industry or closing down polluting companies, are considered as urgent and necessary. Special emphasis is put on the necessity for spreading this ecological awareness so union and other workers' representatives think about the risk of an anti-ecological job, and value the right to work and the right to a healthy environment in the same way. This implies training aimed at changing attitudes assumes ecological complaints on the part of workers. An "ecological literacy campaign" in the words of one of the trades unionists interviewed.

A more continuous and less radical attitude characterises the second focus. In it the environment and safety at work take prime position, although elements of the previous focus are included with the aim of updating its content. "The health-environment term, from a trade union point of view, has to be tackled from the inside of the company where specific problems rise everyday and in the industrial evolution of work centres." (Paper 5th Congress. CC.OO. 1991).

An integral concept of workers' health prevention cannot be confined solely to the prevention of effects which may be caused by work conditions: the environment and safety at work have to transcend these limits and achieve the totality of factors which may influence health, amongst these are the living environment of workers, generally situated close to work and, therefore, influenced by the industrial activity carried out by the company by whom the worker is employed. (Paper presented at 5th Congress CC.OO 1991).

Criticism of productivism by ecological trades unionists is limited, and instead a more reformist attitude is adopted. The restricted position of closure of companies is replaced by emphasis on corrective measures or harsher measures when there is no alternative.

Concern over costs owing to the growing number of environmental regulations, affects unions in different sections of policy in such a way that the policy attempts to balance out different factors to the detriment of more critical attitudes. The following quote refers to this issue.

"The growing concern of industrial societies about economics as a whole, more particularly about ecology and specifically the entry in May into the Single European Market will involve the establishment of more restrictive environmental regulations over a short period, and consequently will compel companies to adopt costly adjustment programmes at their industrial sites. This may lead to survival problems in certain companies, with serious repercussions on the level of industrialisation in our country, and, consequently, on employment". (Paper presented at 5th congress, CC.OO. 1991).

Finally, there is a demand for training of union members and workers, prioritising expert knowledge or training for new jobs related to the environment.

There is often distance between official policy and opinions of the membership of union organisations. Closure of polluting factories is, in this respect, of great testimonial value, because, in view of the fear of job losses, the attitude of workers does not coincide with that of their leaders.

In view of the problems caused by a number of chemical companies in Cartagena (already referred to in previous sections) a member of the chemical federation of CC.OO. expressed his opposition to the frequent closure of polluting industries, "A factory cannot be closed down every time the wind changes, what is necessary are definitive measures for each company" ("El Independiente" 16-4-91).

A workers' representative and member of CC.OO. expressed the same opinion in view of the problem of INQUINOSA "pollution indices found in our blood are not so alarming. INQUINOSA has recognised that all workers have traces of pesticides. It is the same as if in flour was detected in a baker or cement in a builder it is a consequence of our work. Besides, the analyses were not serious, we want the local Aragon authorities to carry out more comprehensive tests on us, along with a detailed medical check-up, and for them to respect our jobs" ("Interviu" 27th May 1991).

The threats which business often employs and which hang over workers, leads the union to a dead end and turns it into a "hostage" compelled to confront a difficult contradiction: continue with their complaints about pollution or content themselves with keeping jobs.

The unantagonistic nature of "pragmatic trade unionism" is owed to the assumption of this problem, which leads to structuring demands towards correction and reforms as regards the state of the environment, trying to liken itself to that producer-citizen, worker-consumer who sees the threat of unemployment in the factory, and outside sees the threat of harmful surroundings.

The dilemma is inherent, therefore, in the heart of the organisation. Demands for extending the ecological offence law, measures against pollution and the consumption of resources would intensify the threat of companies and for this reason they have to be cautiously developed in order to avoid causing greater discord. These precautions, on the part of trades union leaders, are exemplified by the lack of press coverage of accusations in environmental matters.

There is also frequent disagreement between the union movement and the "green" movement. Contact between these two groups has been marked by mutual misunderstandings, by reciprocated ignorance of each others problems and by conflicts in interest. However, contacts have been made in some areas, and ecologists have participated in compiling union programmes in the environmental field. The activities taking place up to now are the result of the efforts of individuals or small groups. Mistakes and disagreements still occur, but this attitude at least goes some way to reducing the distance between these two sections of the movement so that their collaboration will bear fruit in the future.

5.4 Summary

The most insistent and well defined feature of industrial actors' opinions advocate conciliation between protecting the environment and the continuation of the prevailing growth model. Within this framework there are significant differences. Business opinion tends to attach more importance to technology putting aside political solutions. Trades union opinion tends to give more priority to political interventions, on understanding environmentalism as a potential ally.

The dominance of a conciliatory ideology is related to the complex display of attitudes. Contradictions emerge between environment and growth and actual conduct tends to be paradoxical or simply inconsistent. In the majority of cases, it is the economist motive which dominates the management of conflicts.

However the table presented is not static, but dynamic and there are signs of a change. The group of "eco-responsible" employers is a minority but which is tending to increase. There is also a noticeable increase in "ecological" trade unionism.

6   SUMMARY AND CONCLUSIONS.

In an analysis of Spanish regulations regarding the environment, there are numerous standards as a result of the non existence of a General Environmental law controlling and assigning duties and responsibilities. This makes intervention in prevention, correction, follow-up and negotiation difficult. The nearest thing to this law is perhaps the environmental impact regulation, although this is not intended as a unifying law.

Also, law 38/72 on Protection of the Atmospheric Environment (December 1991), indicates that "the criteria of action would be to prepare a general law in defence of the environment". This "postponement" was justified with phrases such as and "the lack of experience in many aspects" of an economic nature and "the necessary dosage of economic measures which have to have an effect on these affairs."

Secondly environmental questions can also be included in matters relevant to collective bargaining, either through company committees, or through specific committees like health and safety and in so doing give the latter more responsibility.

Despite the Spanish Constitution recognising the right to enjoy "a suitable environment for personal development", post-constitutional regulations are fairly scarce on this subject (the only law worth mentioning is the Environmental Impact law and the development regulation). This has resulted in environmental legislation being somewhat obselete and inefficient.

However, despite the limited development of constitutional mandates in these matters, it seems that concern is emerging about public powers, owing perhaps to the enormous deterioration which the environment has undergone. This limited concern materialises in the environmental impact regulation and, above all in the creation of a specific environmental offence (articles 347 and 348 of the Penal Code) which even though it lacks force is an important advance in legislative development in this issue.

The incidence of environmental problems in the more immediate practice of industrial relations has been traced by the examination of a wide sample of collective agreements (almost five hundred state or autonomous agreements). Collective bargaining is very significant in this area and allows what is happening in practice to be examined, and why it is necessary to establish specific agreements or regulations. Environmental problems were hardly found to have entered into collective bargaining (external environmental problems even less). Apparently, such problems are not considered as an essential element of industrial relations.

The path by which it seems to be more feasible to introduce aspects relating to ecological responsibility in collective bargaining, as shown by the state agreement with the chemical industry, consists of giving a wider than normal interpretation of health and safety regulations in the workplace. The steps of this path would be, firstly, to increase the importance given in collective bargaining to questions related with health. Secondly, to promote preventative health measures, which would open the way for internal environment regulations. And finally, to give way to more general environmental considerations, outside the factory framework.

However, in practice there are two obstacles which are socioeconomic and cultural. The frailty and fragmentation of employment markets (a distinguished feature in Spanish economic reality of recent years), with high rates of unemployment and many temporary jobs, places unions in a defensive position, against new problems. Employers want to limit bargaining to those aspects most directly linked to wages and jobs, keeping the management of other industrial dimensions exclusively to themselves.

There are also cultural restraints. The strong medicalisation of occupational health standards (and, in general, of the social perspective of health problems) blocks the development of perspectives based on prevention and on appropriate management of the surroundings. The domination of monetary culture, the commitment of growth ideology, minimises the social perspective of ecological costs.

Less weakness and fragmentation in employment markets, less medicalisation in health management, less monetarism would seem to be the conditions for ecological responsibility to become part of industrial legislation.

Another factor which could result in a change in the situation described by agreements analysed is external pressure. More pressure from the ecological movement and from consumers on one hand, and more political willingness in the development and implementation of environmental laws on the part of political parties or the government would force industrial actors to include this issue in industrial relations. Changes would also be needed to make external pressure effective. Relations between ecologists and employers and between ecologists and trade unionists need to be built, as currently relations between ecologists and industry even though not intense, are frequently openly confrontational.

With unions and the green movement relations tend to be cordial, despite the existence of undeniable conflicts which both sides attempt to minimise. The prevailing attitude of the enforcing authorities is one of tolerance and environmental legislation is not as effectively enforced as it could be.

There has been a perceptible increase in recent years of environmental issues in the ideology and activities of employers' organisations and the unions, even if it has been at fairly basic levels.

The point of interest for employers is the adaptation of Spanish industry to EC environmental legislation. Pressure is put on the authorities to assume the largest part of adjustment costs, both financial charges and investments in infrastructure. Secondly, employers demand more coordination in environmental administration and the creation of more complete data bases on the state of the eco-systems and the impact of economic activities on these. It is suggested that the environment be integrated into management duties of companies, thereby associating it with technological innovation and the creation of an ecologically sound image aimed at consumers and public opinion. Activities are centred fundamentally on offering employers information on environmental regulations and help in management to get monetary subsidies for adjustments.

On the part of the unions, the gradual introduction into their argument of ecological issues such as the danger of the nuclear power industry, desertification, and pollution of the Mediterranean, can be found. These tend to be close to the point of view of ecological movements. However, only on very few occasions do these approaches get to appear on specific demand platforms. It is very rare that trades union practice includes environmental elements even though these are present in the policy.

A distinguishable feature, in policy and practice is mutual ignorance. The spokespersons of employers in this field are the government, other employers and, sometimes, consumers, but very rarely workers and their unions. One example is the 52nd Bulletin of the Circulo de Empresarios (business circle). This is a notable initiative aimed at increasing environmental culture in the employers' world and defining strategies for this purpose. Out of 218 pages filled with speeches of 19 distinguished business leaders, amongst them the most knowledgable people in this field, there is only one reference to trade unions (refering to the safety and hygiene committee of the General Motors factory in Figueruelas.

Trade unions, for their part, regret the limited ecological willingness of the government and denounce companies as being responsible for pollution, but it is very rare that this denouncement is interpreted into specific initiatives, either in protest and mobilisation, or at the negotiating table. For this reason the mention of ecological problems in the Propuesta Sindical Prioritaria, aimed towards negotiation at the highest level with the government and employers' organisations.

It seems that, in this area, the creation of scope for meetings, debates and confrontation if necessary, where industrial agents would be able to exchange their respective views regarding the ecological crisis, would be a step worth considering.

The most insistent and well defined feature of industrial actors opinions is the presence of an ideology of conciliation between productivisim and ecologism. "Economic growth has to be made compatible with protection of the environment" this is the phrase which summarises the social and political consensus. Only the most active minorities of the ecological movement, on one hand, and the most anti-ecological groups from the economic world, on the other, questions this consensus.

However business opinion tends to attach more importance to technology putting aside political solutions (a view which contrasts with the limited investment made up to now in the so-called "eco-industry" sector). It tends to perceive the environment as a mediator of social consensus ("we are all responsible for the state of the environment, we should all collaborate in finding a solution"). Trade union opinion, especially on the left wing, tends to give more priority to political solutions, on understanding conservationism as a potential ally in the fight aganist capitalism.

The dominance of a conciliatory ideology is related with the complex display of attitudes. Given inevitably, that contradictions emerge, actual conduct tends to be paradoxical or simply inconsistent. In the majority of cases, it is the economist motive which dominates the management of conflicts. Thus, a revision of the available information shows that dumping unpurified waste is an almost generalised practice; infringements of environmental regulations abound; incomprehensible information is widespread; the tendency to minimise of ecological costs is very strong; the argument that limits cannot be put on polluting products because it would involve job losses continues to exist and industry's ecological commitment normally refers above all to those businesses in the so-called "green-industry" sector. Expressed in another way ideology preaches harmony between production and the environment but in practice it is taken for granted that the environment can cope with everything.

The distance between opinions and actions can be observed in employers and workers although responsibility is unfairly distributed. Workers have little opportunity to participate in the management of aspects of industrial life, aside from those of wages and jobs, such as technological decisions, the definition of productive processes, or the orientation of investments.

The distinction made in analysing attitudes does not originate from the taking of sides in the internal conflicts of the industry. The attitude of unions could only be put to the test if it held more weight in the decisions relevant to the field studied. The research does not allow conclusion that more intensive union participation would lead quickly to a more ecologically sound industry. Rather there are reasons to suspect that the differences would not be substantial.

Finally, the table presented is not static, but dynamic - there are signs of a change. The most noticeable is that which marks an internal split in the respective cultures of social actors. Thus, business attitudes can be classified into four groups (of greatest to least ecological committment), of which the first, that of "eco-responsible" employers, is a minority but which is tending to increase. And there is noticeable tension between "ecological" and "pragmatic" of trade unionism, within a framework where the second is clearly dominant (but relatively pervious) and where the first one tends to define and assert its stance more than the other.

APPENDIX 1:

Regulations relating to environmental issues.
General guidelines with rules concentrating on the Environment.
- Spanish Constitution 27-XII-1978: articles 45;148.1.9 and 149.1.23.
- Civil code: articles 389 to 391: 1907 and 1908
- Penal code: articles 347a and 348
- General Health Law 44/86 15th April: articles 19; 39; 40.1 and 42. Specific Regulations.
- Royal Legislative Decree 1302/86 28th June, Evaluation of Environmental Impact.
- Royal Decree 1131/88 30th September which develops Royal Decree 1302/86.
- Decree 1131/88 30th November through which the Regulation of Unhealthy, Harmful and Dangerous activities; and by Order 15-III-63 implementation of above Decree.
- Law 38/72 22nd December of Protection of the Atmospheric Environment and decree 833/75 which develops it. Decree 1613/85 1st August which partly modifies the previous one and Royal Decree 717/87 27th May which modifies Decree 833/75.
- Order of 18th October 1976 of Protection of Atmospheric Pollution originating from Industry.
- Decree 3025/74 on Limiting Vehicle Pollution.
- Royal Decree 2024/75 23rd August where the characteristics, qualities and conditions of use of fuel are set out.
- Law 29/85 2nd August: Water Law (articles 2 to 12; 38 to 44; 84 to 103; 105 and 108 to 112). Royal decree 849/86 11th April which develops the Water Law.
- Coastal Law 22/28: articles 1 to 6; 20 to 42; 44; 51 to 65; 72; 85; 88; 90 to 115 and Royal Decree 1471/89 1st December which develops the Coastal Law.
- Royal Decree 258/89 10th March on Tipping dangerous substances from land into sea.
- Law 21/74 27th June on Research and Usage of Hydrocarbons and rules of its development.
- Order of 26-V-76, on Prevention of Marine Pollution caused by spillages from ships and aeroplanes.
- Law 4/89 27th March of Conservation of Green Spaces and Wildlife.
- Law 25/64 29th April on Nuclear Pollution and Royal Decree 2519/82 of 12th August: Regulation on Health Protection against ionizing conditions.
- Law 42/75 19th November on rubbish and solid urban waste.
- Law 20.86 14th May on Toxic and harmful waste. Royal Decree, and Royal Decree 833/88 14th May on execution of this law.
- Royal Decree 2216/85 23rd October: Regulation on declaring new and classified substances packaged and labelled as dangerous substances.
- Royal Decree 199/90 16th February by which the General Environment Secretariat is created within the Public Works Ministry.

REFERENCES:

_____Actas de las Jornadas sobre Sindicalismo y Medio Ambiente, Barcelona, 18-23 June 1.990.

AEDA (Spanish Aerosol Association): Libro Blanco de los aerosoles. Barcelona, June 1990.

ALFONSO MELLADO, C. "Legislación sobre Salud Laboral". Paper presented to the Jornades Sindicals de Salut Laboral, Valencia, 1989.

Area de Política Ambiental de IU: "Pacto de competitividad, propuesta de contenidos ecológicos", Madrid, 1991.

Boix y Ferrando, P.: Sobre salud laboral. Unpublished work, 1.987.

Carcoba, A.: "Paticipación de los trabajadores en la prevención de los riesgos laborales". Gaceta Sindical, Madrid, 1.987.

Carcoba, A.:"Participación de los trabajadores en la mejora de las condiciones de trabajo y salud". Paper presented at the III Jornadas de Salud Laboral y Medio Ambiente de Trabajo, Madrid, 1.989.

CC.OO: III Congreso de la Confederación Sindical de CC.OO., 1983.

CC.OO: IV Congreso de la Confederación Sindical de CC.OO.,1987.

CC.OO. (Area de Ecología y Medio Ambiente): Extracto del informe del Secretario General. Ponencias aprobadas en el V° Congreso de la C.S. de CC.OO., Madrid, December, 1991.

CC.OO.: "Criterios para la negociación colectiva de 1993", Madrid, 1992.

CEOE: Conferencia Empresarial 1992: La empresa española en la nueva Europa. 8: La empresa y el medio ambiente, Madrid, 1992.

CEOE and CEPYME Cumbre de la Industria Española: Resumen de conclusiones. Zaragoza, 1990, 23rd and 24th January.

CEV: Los residuos industriales y la política medioambiental en la Comunidad Valenciana. Valencia, June 1990.

Circulo de Empresarios: Empresa y Medio Ambiente, Bulletin n° 52. December 1990.

Congreso de la Confederación Sindical de CC.OO. del País Valenciano, Xest, 1983, resoluciones.

Congreso de la Confederación Sindical de CC.OO. del País Valenciano, Alacant, 1987, resoluciones.

Departamento Confederal de Ecología y Medio Ambiente de CC.OO.: Resumen de actividades, January - September, 1992.

Díaz, E. and E. Gutiérrez: "Sindicalismo y medio ambiente", ___Gaceta Sindical, March 1990.

Escuela Julián Besteiro-Secretaría Confederal de Formación de UGT: Curso "Salud y Medio Ambiente".

Gutierrez, E: "La acción sindical y el Medio Ambiente". Madrid, Fundación 1° de Mayo, Area de Medio Ambiente, 1991.

I.N.S.H.T.: Encuesta nacional de Condiciones de Trabajo 1.987, Instituto Nacional de Seguridad e Higiene en el Trabajo, 1.988.

Moncada i Lluis, S.: La acción por la salud en el puesto detrabajo en España. European Foundation for the Improvement of Living and Working Conditions, 1.991.

MOPU : Jornadas Internacionales sobre Medio Ambiente: la respuesta sindical. Madrid, MOPU, 1989.

Perez de los Cobos Orihuel, F.: "La Directiva marco sobre medidas de seguridad y salud de los trabajadores en el trabajo y la adaptación del Ordenamiento español". Relaciones laborales, 8 (99-111), 1.991.

Plan Cameral de Medio Ambiente, Consejo Superior de Cámaras de Comercio, Industria y Navegación.

Programa 2000: La economía española a debate. PSOE.

Propuesta Sindical Prioritaria, Information and Publications Secretariat CC.OO. Madrid, October 1989.

Resoluciones sobre Medio Ambiente, XXXV Confederal Congress UGT, Madrid 1990.

Resolución Acción Sindical-Salud Laboral, XXXV Confederal Congress UGT, Madrid, 1990.

Resoluciones sobre Salud Laboral y Medio Ambiente del III Congreso Federal de Químicas y Energía de la UGT, Zaragoza, 1990.

SALA, T (coord): Lecciones de Derecho del Trabajo. Valencia. Tirant lo Blanc, 1.990.

UGT: Circular de Negociación Colectiva, Madrid, 1991.
UGT - Extremadura:Resoluciones UGT Extremadura sobre medio ambiente, Secretaría de Acción Social, 1990.

UGT and CC.OO.:Iniciativa Sindical de Progreso,1991.

Unzeta López, M.: La Salut Laboral: el seu tractament en els convenis col.lectius. Barcelona, Generalitat de Catalunya, 1.987.

INDUSTRIAL RELATIONS AND THE ENVIRONMENT:

UNITED KINGDOM

by

Andrea Oates

Page

TABLE OF CONTENTS

1. INTRODUCTION..181
2. THE LEGAL FRAMEWORK..................................184
2.1 Introduction
2.2 Environmental legislation
    2.2.1 Environmental Protection Act 1990
    2.2.2 Environmental Liability....................186
    2.2.3 Access to environmental information........187
2.3 Legislation on the working environment............188
    2.3.1 The Health and Safety at Work Act
    2.3.2 The Control of Substances Hazardous to Health (CSHH) Regulations........................ 189
2.4 International and European guidelines............ 190
2.5 Proposals for future legislation................. 191

3. VOLUNTARY AGREEMENTS BETWEEN THE INDUSTRIAL ACTORS.. 195
3.1 Introduction
3.2 Agreements on participation in environmental issues

4. POLICY STATEMENTS, DEMANDS AND CAMPAIGNS.......... 203
4.1 Employers' organisations
    4.1.1 Confederation of British Industry (CBI)
    4.1.2 Chemical Industries Association........... 205
4.2 Management and employer response to environmental protection......................... 206
4.3 Workers' Organisations and representatives....... 210
    4.3.1 Trades Union Congress
    4.3.2 National Trade Unions..................... 213

Page

4.4 Attitudes of trade unionists to environmental
    issues... 216

4.5 Interaction with environmental activists and
    community organisations........................ 218

5. CURRENT ENVIRONMENTAL CONFLICTS AND THE INDUSTRIAL
   ACTORS.. 220

5.1 Toxic waste disposal

5.2 Industrial pollution............................ 223

6. SUMMARY AND RECOMMENDATIONS...................... 225

7. REFERENCES....................................... 227

1. INTRODUCTION

Concern over environmental issues began to turn to action in the United Kingdom in the early 1970's. Public awareness about the effects of pollution on the environment increased, and environmental pressure groups including Greenpeace and Friends of the Earth were formed.

Since this time, a series of major disasters both in the UK and abroad, including the explosion at a chemical plant in Flixborough, UK, the poisonous gas leak at the Union Carbide plant in Bhopal, India and the nuclear accident at Chernobyl have brought into focus the industry's potential for ecological devastation.

The late 1980's saw a fundamental shift in British public opinion in favour of issues such as conservation, waste recycling, energy efficiency and sustainability. This was prompted by evidence of international environmental degradation including ozone depletion, an increase in world temperature, and destruction of the rainforests, which attracted widespread media attention. Closer to home there was a series of food scandals which brought widespread criticism of agricultural practices including the use of pesticides and fertilizers which resulted in contamination of food and water supplies by pesticide residues and nitrates.

And a series of well publicised clashes between the UK government and the European Commission over environmental issues -in particular drinking and bathing water quality - increased the perception that the UK was lagging behind other states in tackling these issues.

Public awareness on environmental issues has lead to demands for environmentally friendly products from "green consumers" and hostility to development by industries perceived as bad polluters by local communities.

British industry has largely been forced into reacting to new environmental pressures from consumers and from new and more stringent environmental legislation. Surveys carried out for the Financial Times and the Economist in 1989 and 1990 concluded that the pressures on industry to take action to protect the environment are increasing.

Highly publicised pollution incidents including the Exxon Valdez oil spill, a major Shell pipeline leakage into the Mersey, the public outcry over the attempted import of PCBs from Canada for incineration in Britain have had serious consequences for the companies involved both in terms of cost of cleaning up and the damage to their reputation.

The shift in consumer tastes has been very influential in the United Kingdom in persuading companies to "go green".

However, although not contesting its influence in the short term, many believe that it is a target for exploitation and consumers have become disillusioned with the false environmental claims made by companies marketing their products. There is a more widely held view that the green voter will be more influential in the long term as more and more communities have shown that they are not willing to have industries they perceive as polluters developing next to them.

Pressure is also coming from government as legislation introducing tightening restrictions on pollution is introduced. Controls on emissions of harmful substances, the disposal of toxic waste by incineration and the dumping of chemical and industrial waste in the sea have resulted in increased costs of pollution, which is persuading industry to research and implement clean technology and waste minimisation.

Legal limits on pollutants, tax incentives for cleaner technology and levies on emissions have promoted growth in the green sector: ranging from producing pollution control equipment on cars to recycling rubbish, as has the public's perceived refusal to buy environmentally damaging goods. The British environmental market is estimated to be worth around £4 billion.

However, despite the various pressures, the proportion of UK companies with a corporate environmental policy; or a policy of carrying out environmental auditing remains low. And employers are generally reluctant to involve unions in what has traditionally been considered as a management area as a survey carried out by the Labour Research Department in 1991 shows - only 7% of employers had involved unions in auditing.

Perhaps in part due to the dominant political philosophy during the recent period, and the hostility to trade union organisation, the concern expressed by individuals has not been translated into collective action as far as industrial relations is concerned.

Whilst trade unions have been involved in action to protect the environment since this time, early examples were confined to a number of significant but isolated single issue campaigns which were related to health and safety concerns of the membership of those unions. These included campaigns against the use of hazardous and environmentally damaging substances including pesticides and asbestos, and the refusal by seamen and transport workers to handle nuclear waste destined for disposal in the North Sea.

In the last two years, the TUC Environmental Action Group, which was only formed in 1989 and reported for the first time in 1990, has called for environmental rights for workers, and national unions have mounted high profile campaigns demanding that employers recognise that workers and their representatives have a right to be involved in environmental issues at their workplace.

The second wave of union representation is all embracing, with a shift in emphasis to the collective green agreement which addresses environmental issues at work in a comprehensive and structured way. This approach is in the early stages and there are very few negotiated agreements giving trade unions rights of involvement in environmental issues so far.

The UK trade union movement is conscious that in order to play a more active role in environmental protection, they must challenge the individual and consumerist response and develop it into one of collective action.

2. THE LEGAL FRAMEWORK

2.1. Introduction

There is very limited provision within the formalised legal framework of industrial relations in terms of employee/employer consultation in the UK. There are no legal bodies representing a company's workforce and protective legislation on working conditions is very selective and undergoing revision. Regulatory legislation is concerned with areas including collective action, health and safety at work, and contracts of employment.

Legislation on employee involvement is very underdeveloped in the UK compared to countries such as Germany. There are no rights for the workforce to be consulted and informed about training issues or personnel issues for example. Limited rights of consultation regarding changes in operations at a plant which have health and safety implications are provided within the Health and Safety at Work Act 1974.

Further limited rights to information and consultation are provided in the Transfer of Undertakings regulations and the Employment Protection Act 1975 which both enable unions to receive information on changes planned at the workplace particularly where job losses are envisaged. More recently, the Companies Act of 1985 and the Employment Act of 1982 require companies with more than 250 employees to include in their annual reports a statement describing particular arrangements made to encourage employee involvement.

It is fair to say, however, that in the field of environmental protection no comparable rights to consultation or participation have been extended to representatives of the workforce.

2.2 Environmental legislation
2.2.1 Environmental Protection Act 1990

In line with this the Environmental Protection Act 1990, which began to take effect at the beginning of 1991 and is the most comprehensive piece of environmental legislation introduced in the UK so far, has nothing to say with regard to the role of employees or trade unions.

In considering the impact of the most polluting processes on land, air and sea, it overhauled the existing system of previous piecemeal legislation. However, whereas the Health and Safety at Work etc Act 1974, which similarly overhauled health and safety legislation, created the Health and Safety Commission, (HSC) there is no equivalent body created by the EPA.

The HSC is the national authority responsible for policy making and the formulation of new legislation related to occupational, health and safety in industry and commerce. It includes employer and trade union representation. In comparison there is no national body responsible for environmental protection on which trade unions have representation.

The main sections of the Act extend and strengthen pollution legislation through a new regime of Integrated Pollution Control (IPC). This requires that the "best practicable environmental option" be applied to the control of solid, liquid or gaseous wastes.

The Act also introduces a new system for waste control on land that operates alongside existing hazardous waste legislation. Other sections of the Act cover a multitude of issues, including statutory nuisance and clean air, litter, radioactive waste, genetically modified organisms, dumping at sea, stubble burning, hazardous chemicals and nature conservancy.

IPC is the responsibility of Her Majesty's Inspectorate of Pollution (HMIP). It is limited to around 32 main process categories. A second tier of control created by the Act, Air Pollution Control (APC), falls within the remit of local authority environmental health departments. This will ultimately cover in the region of 25,000 types of installations.

The Environmental Protection (Prescribed Processes and Substances) Regulations 1991 sets out a list of the processes capable of causing pollution which relate to IPC and APC. This covers fuel and power production, metal working, the mineral and chemical industries, waste disposal and a number of other commercial activities.

In carrying on a prescribed process the best available techniques not entailing excessive cost (BATNEEC) must be used to minimise releases into the environment, and to render harmless those substances that are released.

Applications for authorisations are required to implement a combination of processes, pollution controls and treatment techniques that together constitute the best practicable environmental option. This involves adopting techniques that are the most effective in preventing pollution.

All new and substantially varied processes will now be subject to IPC. Existing processes are to be phased in from April 1991 to the end of 1996. In the meantime process operators will continue to operate under their existing authorisations although large combustion plant is already subject to IPC, implementing the requirements of a European directive.

Applications for authorisations from the enforcing authority require the provision of information on various aspects of the operations, and a scheme of charges is applied. There are similar provisions under APC. Applications for authorisations for the first batch of processes had to be submitted between 1 April to 30 September 1991.

Although the EPA does not require employers to carry out environmental audits, the information which has to be supplied for the authorisations, it has been argued, will require an assessment of environmental impact to be made. Employers will also have to show that they can operate the processes and comply with any regulations made limiting emissions and discharges.

Applications to HMIP for authorisations require the provision of information amounting to some ten sheets and describing the process involved, the procedures for dealing with operational breakdowns, staffing, techniques, substances released and concentrations of releases.

Authorisations can only be given where HMIP is satisfied that the conditions it imposes will actually be complied with. These conditions must be suitable for ensuring BATNEEC will be used in carrying out the process: compliance with international law as directed by the Secretary of State for the Environment (SSE); compliance with any quality standards and objectives set by SSE; and compliance with any requirements of the SSE.

Similarly, there is no obligation on employers to appoint environmental co-ordinators or officials responsible for environmental issues; but the introduction of new environmental protection legislation is forcing many large companies to establish such a post.

However, concern has been expressed that there are insufficient numbers of pollution inspectors to be able to enforce the Act effectively. A recent National Audit Office (NAO) report found that polluters in Britain are going unpunished, factories are not being inspected and progress towards integrated pollution control has stalled because the HMIP is seriously understaffed.

The NAO said that the inspectorate needs twice its current staff by 1994 to fulfil its role of monitoring compliance with the EPA. The report also found inconsistencies in decisions to prosecute for air pollution.

2.2.2 Environmental liability

The EPA introduces a new legal duty to exercise "reasonable care" on enterprises which produce waste or transport, dispose or process it. The deposit of controlled waste must be covered by a waste management licence issued by a Waste Regulation Authority.

Breaches of the duty of care can lead to imprisonment. Holders and producers of waste must provide details of the waste and take steps to prevent any illegal dumping.

Greenpeace Business a magazine published by the environmental pressure group aimed at industry, reported that twelve families who have children with respiratory problems are to bring an action for alleged nuisance and negligence against ICI and British Steel in the Teeside area. The solicitor representing the families said that the cases will be seen as a test case for a claim for damages for personal injury caused by environmental pollution from chemical and steel production.

Greenpeace Business said that these are the first in a series of cases to be launched by claimants who believe that exposure to atmospheric pollution over long periods has damaged their health or livelihoods. Environmental protection law also places duties and responsibilities on employees. These concern duties of "care" reporting, and the duty to ensure that instructions given are legal.

2.2.3   Access to environmental information

There is no legal obligation on employers to publish any yearly report on the environmental performance at company or national level. Similarly, companies are under no legal obligations to discuss environmental issues with the workforce, unions, or the local community.

However the EPA requires enforcing authorities to hold public registers of information including documents relating to applications, authorisations, and convictions. Waste Regulation Authorities, for example must provide publicly accessible registers of all relevant documentation.

Inspection of a register by the public must be free of charge and only limited restrictions apply to information deemed commercially confidential. But in practice it is difficult to gain access to information on emissions.

The method by which the data is collected requires the public to visit the local offices of regulatory agencies to obtain pollution details; the National Rivers Authority for discharges into water; the Radiochemical Inspectorate of Her Majesty's Inspectorate of Pollution (HMIP) for radioactive discharges; the local Environmental Health Department and HMIP for discharges into the atmosphere, and the local waste authority for dumping into landfills.

It is expected to take ten years before HMIP have all this information from certain types of processes stored on data bases in its regional offices. And this will only provide monitoring data for a limited range of chemicals through a small number of samples. It will not give total emissions into the environment of toxic chemicals from industry as is required, for example in the United States.

The TUC argue that legal protection against dismissal for "whistleblowing" employees is necessary. There is currently no such protection and an employee who passes on information about a company's poor environmental practice could be dismissed on grounds of a breach of contract. There are implied obligations on the part of employers and employees regarding confidentiality.

2.3   Legislation on the working enviroment
2.3.1 The Health and Safety at Work Act

The Health and Safety at Work (HSW) Act came into force in 1975. It provides comprehensive, integrated system of law dealing with the health, safety and welfare of
people at work, and of people affected by work activities.

Under the Act employers have a general duty to ensure the safety health, and welfare of their employees, to consult them concerning arrangements for joint action on health and safety matters. These consultation and information rights are through the unions not through any separately elected body.

In 1978, the Safety Representatives and Safety Committees Regulations came into effect and gave recognised unions the legal right to appoint workplace safety representatives who have the right to represent their members' interests in relation to any matter affecting their health and safety, make representations to their employer on health, safety and welfare matters and obtain necessary information from their employer to enable them to carry out their functions.

Where two safety representatives request the employer to set up a safety committee, health and safety, the employer is obliged to do this and most health and safety negotiations take place at workplace level through the safety committee. This consists of management representatives, and safety representatives, although there are examples of safety committees at national level.

Safety representatives are entitled to carry out periodic inspections, investigate accidents and dangerous occurrences, and receive time off with pay to carry out their job as safety representatives and to undergo TUC or union approved training.

In comparison, there are no clearly definable and enforceable environmental rights available to workers and their representatives.

For example, there are no legal rights to paid time off for environmental training. Whilst union education courses covering environmental issues are being developed, often as part of health and safety courses; an employer can refuse a safety representative training leave if the course does not exclusively address health and safety matters as prescribed under the 1977 Safety Representatives and Safety Committee Regulations.

2.3.2 Control of Substances Hazardous to Health (COSHH) Regulations

Legislation covering hazardous substances has been introduced within the legal framework on health and safety. The COSHH regulations cover substances classified as very toxic, toxic, harmful, corrosive or irritant; substances which have been assigned a maximum exposure level or an occupational exposure standard, micro-organisms arising from work activities which are harmful, substantial concentrations of dust and other substances which create a comparable hazard.

The regulations require employers to carry out an assessment of the health risks presented by exposure to hazardous substances at work. This involves identifying what hazardous substances are used in the workplace, the risk presented by the use of such substances, how workers may be exposed to them, and what control measures are necessary in order to reduce the risks.

The regulations introduced a hierarchy of controls through which an employer should work in order to prevent, or where prevention is not practicable, adequately control exposure to hazardous substances. Where practicable, exposure should be prevented by elimination, or substitution by a less hazardous substance, or by the same substance in a less hazardous form.

Only where prevention cannot reasonably be achieved should exposure be controlled through measures such as local exhaust ventilation or enclosure of the process. Controlling exposure using personal protective clothing should be a last resort where other methods of control are not practicable.

The regulations require employers to provide information, instruction and training to employees who may be exposed to hazardous substances on the "risks to health created by such exposure and the precautions which should be taken".

Workers are also entitled to know the results of monitoring of the work environment and the collective results of any health surveillance. In contrast, there are no rights for workers to be informed of the results of monitoring of emissions and discharges into the environment. The regulations have resulted in increased access to information on hazardous substances. The Health and Safety Executive (HSE) guidance to the regulations states that "employees or their representatives at the place of work should be informed of the results of the assessment". Again, there is no such guidance with regard to informing employees or their representatives of any environmental impact assessments under the EPA.

2.4   International and European guidelines

At international level there have been developments on industrial relations and the environment. The United Nations Economic Commission for Europe (UNECE) conference took place in Bergen in May 1990. Representatives of governments in Western and Eastern Europe and the US and Canada, and representatives from industry, labour, environmental, scientific and youth organisations took part.

The result was a "Joint Agenda for Action" to which the British Government is a signatory. This aims to promote sustainable development and covers a variety of issues relevant to industrial relations.

These include: the promotion of environmental awareness which includes promoting public participation in decision making and the "active involvement of people in the processes and decisions affecting their lives and environments;

The achievement of clean production including "mechanisms for consultation and co-operation between management and labour" to prevent industrial accidents and hazards;

Drawing up environmental audits on which employers should consult workers and their trade union representatives, and

The provision of full training to workers, their trade union representatives and managers on the environmental aspects of their work.

The 1990 International Labour Organisation Conference passed a resolution on "environment, development, employment and the role of the ILO" which spoke of the need for tripartite co-operation at national and international level to create links between environmental protection and employment creation.

This stressed that the working environment formed an important and integral part of the general environment as a whole and called for integrated policies based on "full collaboration" between, amongst others, employers and trade unions. The resolution also stated that
the restructuring of enterprises to take account of the need for better environmental performance should be promoted while maintaining or increasing, as far as possible the number of jobs.

The European Community (EC) has been the main influence on the introduction of pollution control in the UK, with the UK Government being slow to implement many of the measures, and opposing many of the measures during debate of proposals. Over 280 environmental laws have been introduced since the Community's environmental action programme was adopted in 1973.

In the area of major industrial hazards, the focus on occupational health and safety is being broadened to cover environmental issues. Legislation has been introduced in the UK to implement EC directives to control risks to workers and the wider public from major industrial hazards from, for example, large scale fires and explosions.

Although this legislation was introduced under the health and safety at work framework, it is currently being developed by the EC to address the wider environmental consequences of major industrial accidents as well as disaster counter- measures.

Recently the EC has started to link the issues of industrial relations and the environment. The Commission's 1990 annual report on Employment in Europe contained for the first time a chapter on environment and employment. This is seen as confirmation that there is a "social dimension" to environmental control.

The EC's Structural Funds were identified as being suitable instruments for bringing about change. The distribution of these funds is undertaken on the basis of formal consultation with a number of bodies including trade unions and employers. Participation is a key component of the EC social charter.

The Commission is currently considering the possibility of a draft directive requiring environmental audits to be undertaken and the ETUC is lobbying for any directive to give unions the right to involvement in compulsory audits.

2.5 Proposals for future legislation

The Government's environment white paper was published in September 1990 and although it was claimed to be "Britains first comprehensive survey of all aspects of environmental concern", and contained 350 proposals to tackle a range of issues, it was widely criticised
by unions and green groups for being little more than a review of existing policies. It contained few new proposals and only vague commitments with no clear timetable. (An annual report reviewing progress achieved has subsequently been produced.)

It made no mention of environmental protection at the workplace, of partnership between workers and employers; or to the social and employment aspects of achieving sustainable development.

The new measures announced in the White Paper include:

* appointment of "green" ministers in each Government Department;
* making HMIP an executive agency under the "Next Steps" initiative;
* new measures to combat noise pollution;
* development of new air quality standards;
* increased emphasis on improving energy efficiency, including more funds for the Energy Efficiency Office;
* plans to promote Combined Heat and Power and renewable energy resources;
* higher fines for water pollution;
* tighter local planning controls;
* extension of vehicle MOT to include tuning and emission tests;
* improve speed limit enforcement;
* Government help for historic cathedrals and new woodlands;
* establishment of official eco-labelling scheme;
* publish more statistical data on environmental topics;

Although the white paper briefly refers to support for "staff involvement" in company green policies, no explanation or detail is provided on the contribution citizens can make as employees and members of trade unions. A reference is made to the need for more environmental linked training and its importance, but no new proposals are included.

The TUC pointed to omissions in the white paper including failure to identify the link between the working and living environment; total disregard of the interface between environmental protection and health and safety; and the lack of an attempt to promote worker/unions involvement or to envisage joint union/employer action.

It is unlikely that the current Government will introduce legislation which will lead to greater participation of unions in workplace environmental issues; particularly as they are seeking to restrict all other areas of union activity.

The Labour Party's proposals are outlined in "An earthly chance", Labour's programme on environmental issues. This commits a future Labour Government to extending environmental rights at work as proposed by the TUC, and to establishing an environment commission.

The Trade Union Congress (TUC) have proposed that unions should be involved in environmental issues at the workplace, and says it will campaign for a change in the law if companies are unresponsive to the proposals, which include rights to consultation and training (as is currently the case for health and safety) and rights to information based on a process of joint union/employer environmental auditing.

The Environment Commission would be established on a similar basis to the Health and Safety Commission (HSC), which is the main policy making body on health and safety in the UK and enforces health and safety through its executive (HSE). It would deal with research and policy, and would have an independent chair and membership drawn from government, industry, trade unions and representatives of the environmental movement.

An Environmental Protection Agency would be established which will be independent of Government. This would set new pollution standards, timetables for achieving standards and operate under a new freedom of information regime.

The Labour Party says that a Freedom of Information Act will be an early priority for the next Labour Government. This will set up and maintain registers of environmental information, presented in an understandable form, covering water and land contamination, pollution permits, food quality and air pollution and providing raw data rather than smoothed out averages. Registers would also be available in on-line electronic form for ease of access by campaign groups, voluntary bodies and research groups.

The Labour Party has also endorsed proposals for a new set of regulations to control environmentally - damaging substances (COEDS) put forward by the Communications Union, NCU. The proposed regulations would be made under the Health and Safety at Work Act and work on a similar basis to the existing Control of Substances Hazardous to Health (COSHH) Regulations.

The regulations would oblige manufacturers and suppliers of substances deemed to be environmentally damaging to declare what these substances were comprised of, how they should be used, and how they should be disposed of in an environmentally sensitive manner. Breaches of the regulations would expose the guilty parties to punitive fines along the lines of the "polluter pays principle."

The proposal also include the convening of a tri-partite body by the Health and Safety Executive's Policy Division to make recommendations to the Commission on what substances are covered by the Regulations and which are not.

A code of practice would be issued under the regulations which would require the disclosure of key information by manufacturers and suppliers of designated environmentally damaging substances to anyone who requests them but also, as a matter of obligation, to anyone who could reasonably expect to come into contact with the substance by means of purchase, transportation or use at work or in the workplace.

The information to be disclosed would include what the environmentally damaging substance is, what its chemical, biological or physical composition is, what facilities for safe handling and storage are required, what the hazards of the substance are likely to be if misused or corrupted, what remedial action would be necessary to make safe and keep safe the substance, what the approved uses of the substance are, and how the substance should be disposed of. This information would be provided in comprehensive data sheets.

3. VOLUNTARY AGREEMENTS BETWEEN THE INDUSTRIAL ACTORS

3.1 Introduction

The United Kingdom has no written constitution, and therefore no entrenchment of fundamental rights and freedoms. This applies in the industrial relations field as well as elsewhere. There is no automatic right of recognition of trade unions in the UK. A trade union is "recognised" for statutory purposes if it is recognised to any extent by the employer for the purposes of collective bargaining. The Trade Union and Labour Relations Act 1974 sets out the matters which the law considers may be the subject of collective bargaining.

Collective bargaining is carried out on a voluntarist basis. Collective agreements are at present not legally binding on either side, although there is a presumption that those items which are capable of being incorporated into an individual contract are legally enforceable. Procedure type agreements are not legally binding and are therefore legally unenforceable.

Collective agreements are the end product of collective bargaining which may be conducted between an individual employer and a union, a number of employers and a union, or a number of employers and a group of unions. It is common for more than one union to be recognized for the purposes of collective bargaining within a company, or factory and so on.

The collective agreement is the key factor determining wages and other conditions. Company agreements or agreements by occupational groups are of most importance. There is a continuing trend towards company agreements, although there are still 200 permanent industry-wide negotiating bodies in the UK known under a variety of names such as National Joint Councils (NJCs), Joint Industrial Councils (JICs) and National Joint Industrial Councils (NJICs.) Half the members are appointed by the trade unions and the other half by management.

There are some employer federations which negotiate with appropriate unions at industry wide level, but national bargaining has broken down recently in a number of sectors, notably banking and food retailing, so national bargaining is now the exception rather than the rule. The majority of agreements in the private sector are concluded at company level.

3.2 Agreements on participation in environmental issues

The involvement of the trade unions with environmental issues has developed over the last twenty years or so. Early action tended to be very reactive and concentrated into significant, but isolated areas of industrial action and campaigning. Environmental bargaining has tended to be ad hoc and subject specific, reflected in a series of successful union campaigns promoting the use of non-toxic chemicals, recycling, and the use of lead free fuel.

The collective green agreement has only recently been sought by unions with the second wave of union representation which began with the first report of the TUC environmental action group presented to Congress in September 1990. In this approach the emphasis has shifted towards all embracing bargaining and green agreements which address environmental issues at work in a comprehensive and structured way.

The "model green agreement" outlined by the TUC is typical of green agreements. which unions, including the general union, the General Municipal and Boilermakers Union (GMB) and Manufacturing, Science, Finance Union (MSF) are attempting to negotiate with employers.

Full text of the Model Green agreement

"The union and the employer recognise the increasing impact of environmental issues on the employer's operations.

They agree that the need to meet higher environmental standards also presents an opportunity to achieve better levels of environmental quality and business efficiency.

The responsibility to achieve higher environmental standards rests with the employer. This requires the employer to make significant improvements in the quality of its environmental management and to develop new skills in the workforce. This is best achieved by the active cooperation of the union.

Such an approach can only succeed if there is a clear, joint strategy that spells out both the challenges for the employer and the ways in which workers and managers will respond to these challenges.

The union's input into this strategy will be via the joint Health and Safety Committees (or any other joint body if agreed by both parties).

This agreement lays down the framework within which the strategy will be developed.

Joint environmental policy

Any environmental strategy will need to be flexible enough to cover different challenges within all parts of the organisation. However, the signatories believe that the following issues will need to be addressed:

a) the environmental impact of the employer's current and proposed activities, ranging from the workplace and its immediate surroundings to the international level. Consideration will need to be given to:

- waste or by-products (liquid, sold or gaseous) from the employer's operations.

- products and raw materials used by the employer in the course of its activities.

- disposal, re-use and recycling of products when they have ended their initial "working life".

b) challenges and threats to the employer's activities from legal, social, commercial, economic and other similar developments.

c) opportunities for the employer to develop new products processes and services, as well as competitors' efforts to exploit similar opportunities.

d) action by the employer or the union to influence and anticipate public debate on environmental issues.

Joint examination of the Issues

The signatories agree that regular, structured communications and discussions between management and union representatives are essential to the success of a joint environmental strategy.

These discussions will reassure and inspire employees and provide management and union representatives with insights that are essential to the success of a joint environmental strategy.

They should take place through the official joint Health and Safety Committee, or any other joint body as agreed by both signatories.

Both signatories can place any relevant issue on the agenda and have that issue fully discussed, without in any way committing either side to support a particular idea.

In order to gain the most from these discussions it will be necessary to make use of a range of expert advice, for example, employer or trade union federations, environmental consultants, academics and public agencies, both local and national.

Some aspects of the employer's activities may be commercially or otherwise sensitive and therefore the discussion of issues or the supply of data will sometimes need to be governed by principles of confidentiality.

However, both parties believe that the test applied to any question of disclosure should not be "the need to know" but "the need not to know". Disclosure of information is a good thing and information should not be withheld simply because of possible "embarrassment".

Environmental audits and impact assessments

The parties agree that decisions about environmental action must be made on the basis of hard facts, not emotion.

They agree that both internal and external audits of existing operations and impact assessments of proposed operations are an essential tool in strategic planning.

Both sides accept that such studies, although sometimes involving the commitment of considerable short term resources, tend to more than pay for themselves in the medium or long term. They also accept that independent verification and expertise may sometimes benefit these exercises.

The parties believe that audits and assessments will encourage managers and union representatives to take a longer term view of all aspects of the employer's activities, leading to significantly improved all-round performance.

They also agree that audits and impact assessments will benefit from the maximum possible input from and openness to, local authorities, community and environmental groups and regulators.

Training and education

It is vital that union and employer representatives have a good working knowledge of environmental issues.

The employer therefore agrees to cooperate with attempts by union representatives to gain knowledge and receive training about the complex issues under discussion, and to ensure that its managers also receive appropriate training.

Such cooperation may include:

* paid time off for union representatives to attend educational courses;
* funding of training for both sides representatives;

- working with experts to advise both signatories representatives and to inform the workforce;
- producing material for employees which outlines "green" challenges and opportunities for the employer;
- developing an information campaign aimed at persuading the local community, the media and regulatory authorities that the employer is taking seriously its environmental responsibilities.

This agreement will run until amended by the agreement of both signatories or revoked in writing by one signatory with six months written notice."

Bargaining initiatives have been launched by three of the large UK unions. The general union, GMB, which represents around 900 000 members across a wide range of industries, and whose general secretary chairs the TUC environmental action group, launched its "Green Works" initiative in 1991. Twenty five companies have been identified, including local authorities, and chemical companies, which the union wants to sign its model environmental agreement.

The agreement contains provisions for joint union/employer environmental auditing, training for union representatives on environmental issues, and the provision of information to employees, the local community and the media.

As part of the initiative, negotiators in all the union's industrial sectors will be issued with checklists providing a step by step guide to "greening" the workplace. Union representatives are encouraged to raise environmental issues with the membership.

The initiative recognises that although people are generally more aware of environmental issues, and many are involved in campaigns to protect the environment, trade unionists do not necessarily see the union as a vehicle for their concerns.

MSF, which has 650,000 members in manufacturing and the public sector and which represents workers in the chemicals industry has approached chemical companies asking for initial talks on involvement of the unions in environmental affairs, whilst producing an information pack for negotiators. The union's annual pay claim to ICI in 1991, which was negotiated at national level with the company, included a request for the setting up on an environmental committee.

Again a model environmental agreement has been produced which includes the setting up of an "environmental committee" to act as a forum for discussion between the company and MSF on environmental matters. This contains provisions for the environment committee to be informed on environmental protection investment, development of clean technologies, pollution and emissions, health and safety issues, environmental auditing activity, compliance with environmental legislation, storage and transport of hazardous materials, waste management and environmental education.

The communication union, NCU, which represents some 150,000 communication workers, has launched a green code for its union officials. This green code includes the discussion of environmental issues at meetings, rudimentary environmental audits as part of safety inspections, measures for ensuring enforcement of environmentally sound measures, programmes for boosting awareness amongst members, and suggestions on how environmental work can be made more effective by linking up with local pressure groups. This is intended to provide a framework for more detailed negotiations at local level.

The union has also reached an agreement with British Telecom (BT) for the joint monitoring and discussion of environmental initiatives launched by the company. This includes a joint forum for the union and the company to give impetus to the initiatives being put forward by both organisations.

However, there are major difficulties facing unions attempting to play an active role in environmental protection and it is clear that national policies and initiatives have not yet been translated into collective action at local level.

There is currently no requirement for companies to carry out environmental auditing, which has been identified by the TUC as the main area where unions can be involved in environmental protection at the workplace. Many organisations in the UK have yet to assess their environmental performance, or have not yet developed an environmental policy.

And even where companies do carry out auditing the tendency so far has been to do this without the involvement of the unions. A survey by the Labour Research Department in 1991 which looked at the involvement of unions in environmental issues at workplace level found that only 17% of employers in the survey had carried out an audit, and only 7% had carried out joint management/union audits. A survey carried out by Ruth Hillery in November 1991 looking at corporate environmental management attitudes found that 12% of companies undertaking environmental audits include and similarly, the lack of legal rights of union consultation and negotiation is a serious difficulty since it is unlikely that employers will enter into negotiations on environmental issues, often seen as being sensitive whilst they are under no legal obligation to do so.

Although the Labour Research Department survey found that almost half the employers negotiate with unions on environmental issues, the survey was fairly heavily weighted towards the public sector practice as 74% of responses came from union representatives in the public sector and 26% from the private sector. Also, the survey only covered unionised workplaces and as such is likely to reflect best practice rather than give an overall picture of industrial relations and environmental issues.

The survey found that over half the negotiations took place at the safety committee, which suggests that the negotiations were mainly on health and safety issues where there are legal rights for union involvement. Only a very small number of employers (less than 2%) had set up an environmental committee where negotiations with the unions on environmental issues took place.

Lack of access to information is also a serious difficulty. Employers are currently under no obligation to give workers or unions representatives access to information on any aspect of the environmental impact of the workplace, apart from information on hazardous substances under the COSHH regulations.

The Labour Research Department survey showed that as well as a very low number of employers carrying out joint environmental auditing, only 3% had provided training on environmental issues. This suggests that employers are reluctant to give information to trade unionists on the environmental impact of the company, or provide training and education to encourage workers to become involved in environmental issues at the workplace.

However, in spite of the difficulties outlined, there are examples of joint action and co-operation between employers and unions on environmental issues. The Labour Research Department survey found six examples of environmental committees set up in workplaces where management and unions discussed environmental issues. These were all in public sector workplaces. The survey also found 23 workplaces in the private and public sectors where joint management/union environmental auditing is carried out and 12 employers providing training and education on environmental issues.

A number of examples of joint action and co-operation are outlined below:

West Wiltshire District Council (a local authority) discusses environmental issues at the local joint consultative committee, along with equal opportunities, leave and health issues. An environmental strategy was drawn up in February 1991, after consultation with the unions, and as a result recycling schemes have been set up in the area.

At Dunlop Ltd, which manufactures rubber products, environmental issues not related to safety are discussed along with safety-related issues at the safety committee. Recycling of waste has been discussed with GMB representatives and a scheme has been implemented.

Energy efficiency initiatives, including insulation and improved heating systems were initially raised at the safety committee by Institute of Professionals, Managers, and Specialists (IPMS) safety representatives at the Natural Environmental Research Council. These issues were then discussed at the local Whitley committee (joint consultative committee) before being taken up by management.

A waste recycling working group was set up as a sub committee of the joint consultative committee at Sheffield Polytechnic to examine the implementation of a higher grade paper collection and recycling scheme.

There are also examples of workplaces where green groups and community group had been involved in the implementation of environmental improvements.
At SCM Chemicals in Humberside, a chemicals plant producing titanium dioxide pigments for paint, the joint shop stewards committee, comprising of representatives including the Transport and General Workers Union and general union GMB successfully negotiated an improvement in the company's waste disposal methods. During the negotiations the unions contacted Greenpeace and arranged a meeting with senior managers, the green group and the unions to discuss solutions to the problem of discharge of chemical waste into the Humber Estuary.

And at another company, Amerlite diagnostics, local site liaison committees are held with local community groups.

4. POLICY STATEMENTS, DEMANDS AND CAMPAIGNS

4.1 Employers' organisations

4.1.1 Confederation of British Industry (CBI)

The CBI is the central organisation of employers in the UK and seeks to represent the interests of its members at national and international level. The CBI is not a negotiating body, but seeks to formulate and influence policy on industrial and economic matters.

In 1986, the CBI issued guidance under the slogan, "Clean up - its good for business", which outlined how good environmental practice could be profitable. Then, as environmental issues leapt to the top of the political and economic agenda in 1989, the CBI took a decision to take a greater leadership role and actively promote better understanding of the environment and more consistent high performances throughout British business. An Action Plan for the 1990's was drawn up, "Environment means Business" and an environmental management unit was established in March 1990.

The unit aims to take a more pro-active role and encourage business to make positive environmental improvements before being forced to do so by government legislation. Its main role is to promote good environmental practice through the provision of guidance and promotional material on key environmental issues and to establish links with other expert groups.

The CBI a long established environment policy unit, which has a number of specific panels and working parties on which CBI members sit and can help formulate policy on various environmental issues. The purpose of the panels is to monitor, review and shape the development of legislative proposals and enforcement of regulations. The CBI decided that the key issues for priority attention should be human-induced climatic change, ("the greenhouse effect"), waste management, recycling and the tidiness of the UK.

On these issues it has provided information explaining to business the significance of the greenhouse effect and what can be done to alleviate it, and in particular reporting on the role that energy conservation can play. It has also provided general guidance on waste management, in particular how business shall implement a "duty of care", on recycling and on appropriate action to tidy up Britain.

It advises that companies need a sound environmental management system, accompanied by a thorough review or audit of the company's activities which effect the environment, and a written statement showing the company's policy on the environment.

But it says that corporate environmental performance strategies will only be successful with the commitment and support of senior management from board level and the involvement of all employees. And in common with issues such as quality, it says that enhanced staff morale and higher productivity can be the direct results of environmental improvement measures within the workplace where employees are encouraged to become directly involved both during the execution and follow up and advocates suitable education and training schemes at work.

Action which the CBI say companies could be taking is summarised as follows:

* ensuring adequate communication with employees in all aspects of the environmental performance of the company, particularly those areas regarded as sensitive;

* making the environment one of the key issues for discussion at existing consultative meetings and if necessary setting up a specific committee to address environmental issues, with representatives drawn from all functions of business;

* initiating a suggestion scheme on environmental issues and where appropriate "reward" personnel whose ideas are adopted;

* ensuring that employees of all departments of the company are involved in the continuous process of environmental reviewing from the start. The commitment will contribute towards the successful outcome of improvement plans which the process is designed to generate.

It recommends a far reaching and voluntary review of operations and products from an internal point of view, backed up or "validated" where appropriate by an independent body and then published as an authoritative document for internal and external consumption. This, it says, should be a continuous internal process which measures performance against a set of objectives identified by the company.

However, the CBI is wholly opposed to mandatory environmental auditing and to the involvement of the unions in auditing on the grounds that auditing should be a management tool for assessment and control tailored to the circumstances of the company in question. This approach means that some companies may well believe that there should be a formal role for the trade union representatives in auditing whilst others would regard it as a management task. The CBI does not believe that legislation is appropriate to deal with the process of environmental reviewing.

The CBI recognises that there is a great deal of public pressure for more information on the environmental performance of companies as well as for greater openness of information. But sees the real question for companies is to what extent new information should lead to a distinctive new role for the trade unions in decision making.

It is opposed to trade unions playing a more active role in environmental solutions and hence enhancing their involvement in company decision making. It says that there is little doubt that if that role is gained, the trade unions will have achieved their long- standing industrial relations objective of a wider collective bargaining agenda as well as involvement in companies' investment decision.

The CBI also says that environmental issues should be tackled by groups, including businesses, environmental groups, government and the public working together. It believes that a non-adversorial method of working is necessary in order to identify problems and find solutions.

4.1.2 Chemical Industries Association

The CIA represents most UK chemical manufacturers. The chemical industry has been under much pressure since the new wave of environmental awareness. It has agreed a systematic programme of measuring companies' environmental performance and publishing the results. The environmental monitoring programme is an extension of the industry's responsible care programme which was relaunched in March 1991.

All but three of the association's 215 members accepted the programme. The performance indicators include an "environmental index" comprising the five most important parameters for each chemical plant; output of hazardous substances defined by the government as special wastes; output of pollutants which are especially harmful to the aquatic environment; incidents while transporting and distributing chemicals; energy consumed per tonne of product, and the number of complaints from the public.

However, the chemicals industry is opposing moves within the European Community to introduce compulsory environmental audits with the results made public. A "Green Charter" launched by the International Chamber of Commerce and signed by the CBI and the CIA is voluntary and relies on self policing.

4.2 Management and employer response to environmental protection

Businesses are coming under increasing pressure to pursue environmentally sound practices. To ensure that this is publicised, having a "green label" is seen as increasingly important.

Environmental challenges to industry are coming from legislation, consumer pressure, investor pressure and employee pressure. Integrated pollution control has fundamentally changed the basis of pollution regulations, although there are doubts about its effective implementation. Whatever the British attitude to regulation, the European Community and the wider international community are set to be a continuing source of mandatory, harmonising legislation.

There is also consumer pressure, with recent polling evidence suggesting that the market for environmentally friendly products has remained fairly constant over the last year. There is pressure from investors, as those providing finance for business are increasingly questioning whether their investments are environmentally sound. And there is employee pressure as workers become more environmentally aware.

An increasing number of UK companies have produced corporate environmental policies and are conducting environmental reviews of their performance. The consequences of environmental pollution can be disastrous for companies both from a financial and public relations viewpoint.

The recent prosecution and record fine of £1 million for Shell for its pollution of the River Mersey, and the accidents suffered by Union Carbide and Sandoz have been cited by the directors and managers of other companies as the main impetus for going beyond mere compliance.

However the proportion of companies with environmental policies remains low, and the policies are not always put into practice. This is confirmed in a number of surveys which have examined the response of managers and directors of UK companies to increasing pressures on environmental issues.

A survey carried out in 1990 by KPH Marketing of 107 managers from the UK's top companies showed that whilst interest in corporate environmental matters has increased sharply, some 36% of the companies surveyed had not introduced environmental policies.

The main inducement for companies to adopt green policies is said to come from consumers, with employees second and trade unions least influential. The survey also showed that employers are finding that their attitude to the environment is fast becoming an important factor in the recruitment and retention of staff.

An Institute of Directors survey of 500 British directors showed that although directors are more environmentally aware than ever before, many are uninformed about their personal responsibility and legal liability for protecting the environment. The survey showed that few company directors are taking any action to improve environmental performance. Most are said to be paying only lip service to environmental protection.

This was also found to be the case amongst managers in a survey by the British Institute of Management, the results of which were published in 1991. This showed that 52% of organisations which responded to the survey did not have an environmental policy.

While more than three quarters had over the last five years introduced environmental initiatives such as energy conservation, recycling, there were few audits and little staff training. Some organisations believed they had no effect on the environment at all.

The Labour Research Department survey showed that there was a high proportion of employers taking some form of measures to improve environmental performance, reflecting a high degree of awareness. However an analysis of the types of measures taken by employers tended to be in areas related to health and safety, for example the substitution of hazardous substances, or in areas not requiring major changes, for example, purchasing environmentally friendly materials, providing recycling facilities and improving energy conservation. The survey found much less activity in areas involving major changes to the processes.

A CBI survey of 250 firms showed that environmental awareness is increasing, though less so amongst smaller firms. Of those surveyed 35% regarded environmental issues as "very important", but half did not have specific management structures for dealing with environmental issues.

Just under half said they had or were planning to undertake major capital investment on green matters. Some 65% said that local authorities would have the greatest impact in driving their attempts to reduce air and water pollution and the disposal of hazardous wastes. The survey also suggests that management did not perceive the workforce as a major influence on organisational change.

Cranfield School of Management carried out a survey which indicated that small companies do not really regard environmental issues as their problem. Only 2% of small companies surveyed felt they produced a "lot of pollution" and nearly half said they took almost no measures to protect the environment. Less than a third of non-manufacturing companies saw the environment as "very important", and only 14% of manufacturing companies had carried out environmental audits.

There appears to be a reluctance on the part of management to involve workers and their representatives in environmental issues.

The CBI published a paper on corporate environmental policies and environmental statements of 19 companies in May 1991. Of these, only two of the companies outline their environmental policies in relation to employees.

James Cropper plc, a paper manufacturing company based in Cumbria in the North of England, set up an environmental audit group in 1990 to survey its industrial site, and established an environmental policy in which it states that the company will inform suppliers, customers, the work force and the public about the measures taken to protect the environment and seek their cooperation in meeting the company's objectives.

Proctor and Gamble's environmental policy includes assurances that the company will ensure that products, packaging and operations are safe for employers, consumers, and the environment, and that consumers, customers employees, communities and public interest groups and others will be provided with relevant and appropriate factual information about the environmental quality of Proctor and Gamble products, packaging and operations.

"Changing Corporate Values", a report on the social and environmental policy and practice in Britain's top companies points out that it is difficult to assess companies' environmental record in the UK because of the lack of independent information available to the public. It is therefore difficult to assess whether companies which have a corporate environmental policy are putting the policies into practice.

Information availability on a company's environmental record is further limited because the HMIP have a policy of constructive engagement with companies rather than fining them for pollution offences. Also inspection activity by the HMIP is low, with fewer than 500 visits by HMIP inspectors to registered works in 1988/89, which represents a small proportion of the total number of workplaces.

Although companies are claiming commitment to environmental protection in the corporate environmental policies, environmental pressure groups, and the companys prosecution record, have shown that this is not always the case. For example, Proctor and Gamble's example of a lifetime environmental impact assessment of one of the company's products is highlighted in the CBI paper.

The company claim "this was a cradle to grave assessment of products and processes from raw materials, production to disposal of the end product for the "Pampers" disposable nappy product. In studying the environmental impact of Pampers versus traditional towelling nappies, it was shown that there was nothing to choose between the two types in terms of overall environmental impact, when all factors were taken into account".

The research carried out for Proctor and Gamble was questioned by the Womens Environmental Network which also issued a formal complaint to the Advertising Standards Authority claiming regarding the claims made by the company in leaflets and adverts.

Cleanaway is signatory to the Chemical Industries Association "Responsible Care Programme" and has an environmental policy which is again outlined in the CBI paper; but is one of the companies whose polluting activities in the Mersey basin area are highlighted by Greenpeace for consistently dumping excess effluent into the sea.

Greenpeace launched "Greenpeace Business", a bimonthly newsletter in June 1991 which outlines that the organisation has a track record of exposing environmental abuse, and intends to expand such activities into the realm of company practice, making use of the legal system, shareholder and investor pressure. The newsletter says it will highlight sound industrial practice and expose activities of companies which harm the environment.

A survey of the top 1000 UK companies by a political consultancy specialising in environmental legislation and an engineering consultancy found that industry believes that "green campaigners" are having too much influence in drawing up EC environmental regulations. Forty nine percent of the companies said that forthcoming European legislation is biased too much towards the environmentalists' viewpoint.

4.3 Workers' Organisations and Representatives

4.3.1 Trade Union Congress

The TUC is the central co-ordinating body for the trade union movement and is composed of individually affiliated trade unions. Around 90% of all union members, representing around 40% of the working population in the UK, belong to TUC affiliated unions.

In 1972, the TUC organised a conference on "Workers and the Environment" when it acknowledged that "trade unions concern necessarily extends beyond the boundaries of the factory to the domestic and natural environments".

The TUC naturally has the difficulty of reconciling interests of all the different groups of workers it represents. This is reflected by debates within Congress on energy policy, with regard to the acceptability of nuclear energy.

Amongst the TUC's affiliates are unions representing workers in the nuclear power industry, and in the coal and oil sectors. Whilst the National Union of Mineworkers point to the poor safety record of the nuclear power industry, unions representing nuclear power workers point to the burning of fossil fuels as the cause of global warming. TUC policy is in favour of a phasing out of nuclear energy over 15 years by a future Labour Government. An attempt to overturn this commitment at the 1991 Congress was defeated by 4.59 million votes to 3.213 million.

In recent years, particularly since the establishment of the environmental action group (EAG) in 1989, the TUC has concentrated on the involvement of unions at workplace level negotiations. In 1990, the first report by the group contained an overview of environmental issues faced by working people, the proposed role of trade unions, and made proposals for "Greening the Workplace". This included demands for trade unions to be consulted on environmental audits and for extended rights for trade union representatives.

The EAG report outlined that the working and living environment are linked and that most external environmental concerns originate in the workplace, therefore trade unions should have the right to participate in decision making.

It proposes that the role of safety representatives should extend to cover environmental issues, and outlines new environmental rights introduced under which workers would have the right to:

- know the environmental impact of the products and processes they are using and producing;

- be informed and consulted on the environmental strategies of their company;
- initiate and participate in environmental policies, audits and inspections, and a legal entitlement to call in independent inspectors;
- negotiate changes in production and work organisation;
- refuse, without recrimination, to undertake work with potentially harmful environmental effects;
- paid time off to receive necessary education, training and retraining.

The TUC have identified auditing as the main area where unions can become involved in environmental issues at the workplace. It say that audits should cover the use of raw materials and energy consumption, harmful emissions during the production process, waste generation and disposal, the substances and materials used throughout the workplace, and the environmental impact of the product itself. The TUC believe that environmental auditing should be carried out in consultation with union representatives, and that information for this purpose should be made available.

The EAG submitted its views to the National Economic Development Council in 1991 acknowledging that both sides of industry must work together to achieve higher standards of environmental performance.

The EAG believes that Government must play a more supportive role and work in partnership with industry, local government and unions. The group says that there should be an integration of environmental policy with industrial and employment policies in an open spirit of partnership and co-operation.

The group has called for a national environment education and training programme; but also considered that there was an urgent need for trade unions to be better informed of workplace environmental issues and given practical advice on what was required to raise standards.

The group therefore produced a "greening the workplace" guide which provides a "working kit" for unions wanting to build on their environmental activities. This outlines that it believes a consensual approach is necessary to deal with environmental issues and that the union response "extends to an awareness of the possible impact that environmental adjustments will have on wage negotiations, on training needs and workloads".

The guide aims to assist unions bring environmental issues on to the bargaining agenda. The TUC believe that this can result in potential opportunities to discuss and establish union legitimacy in related issues, facilitate closer links with the local community and other outside groups, and make a contribution to the unions image and recruitment.

The TUC has called on all its affiliated members to:

* ensure that trade union policies and activities fully reflect environmental considerations and priorities;

* seek to achieve higher environmental and health and safety standards at the place of work by way of co-operation and partnership with the employer and outside organisations;

* develop joint environmental strategies (by way of a "green agreement" or otherwise) that spell out both the challenges and responsibilities of the employer and trade union;

* agree a framework for implementing an environmental strategy which provides for access to information and regular, structured communications and discussions between management and union representatives;

* assess and evaluate the environmental performance and impact of workplace activities by undertaking jointly agreed environmental audits or reviews;

* raise awareness of environmental issues and the implications both inside and outside the workplace;

* promote the active participation of all employees in joint action to improve environmental performance;

* publicise their activities on environmental issues to the workforce, media and local community;

* ensure that the environmental education and training needs of the workforce are adequately met;

* seek paid time off for union representatives to attend education and training courses on environmental issues.

It is worth noting that the TUC began running Environmental Education Courses as part of its National Education Programme in October 1992.

4.3.2 National trade unions

There has recently been a period of political hostility to union activity on any level, which has resulted in a loss of power of the trade unions. New laws have been introduced on taking industrial action including restrictions on secondary picketing, and balloting members.

Perhaps not surprisingly, there has been a decline in membership of trade unions. According to the Government's 1991 Labour Force Survey the proportion of British workers in trade unions and staff associations had declined to 37%, from 38% in 1990 and 39% in 1989. One of the reasons why unions are seeking to extend their role into the area of environmental protection is to re-establish their legitimacy and increase recruitment, and to extend their role into related issues such as investment and training.

Trade unions are increasingly taking the view that environmental issues are likely to remain on the political agenda and changes in the way industry operates are inevitable. The unions want to ensure that their members are involved in progressing the changes.

Recent years have seen trade union campaigns on the environment gain momentum. Most of the largest twelve unions in particular GMB, MSF, NCU, TGWU and NALGO have been particularly active in this area. These have launched initiatives to translate national policy into action at
workplace level and put environmental issues firmly on the bargaining agenda.

These campaigns have tended to be run along two lines. High profile approaches have been made to employers demanding that the union's role in environmental issues is recognised, and that employers negotiate with unions, carry out joint environmental auditing, and provide access to information on the company's environmental impact. At the same time, training and information has been provided in order to educate the membership in order for them to be able to negotiate at workplace level.

There has been a change in emphasis from campaigning on particular environmental issues, towards demanding all embracing green agreements, reflecting TUC policy.

Within the last few years, following an increased level of awareness amongst the membership of unions which has resulted in the adoption of resolutions on environmental issues unions are beginning to adopt comprehensive environmental policies.

Because of the reactive nature of early examples of industrial action taken by their members on environmental issues, national unions have found themselves in the dilemma of attempting to support workers on opposing sides of conflicts between environmental protection and job preservation.

An example of this occurred in 1989 when dock workers, members of the Transport and General Workers' Union, TGWU, refused to handle imported toxic waste destined for incineration at a waste incineration plant in Wales. The union also represented workers at the incineration plant, situated in an area of high unemployment, and were therefore caught in the classic conflict of representing workers taking action to protect the environment, and workers opposed to the action on the grounds of job preservation.

The TGWU have since made clear its position on toxic waste importation: "The TGWU says that positive immediate action is needed to eliminate this trade in which profits are being put above the safety of workers, communities and the environment. The TGWU says no to Britain being used as the world's toxic waste dump".

The TGWU has also formed an environmental policy group, ENACTS, to look at policy across its trade groups.

The NCU held a conference in April 1991 to launch a new code of practice for union members and to launch its campaign for new regulations to control environmentally damaging substances. The green code for union officials provides a framework for more detailed negotiations at local level. This covers discussion of environmental issues at meetings, rudimentary environmental audits as part of safety inspections, measures for enforcing environmentally safe measures, linking with local pressure groups and raising awareness of the membership.

The union believes that environmental issues are inseparable from health and safety issues. It sees that environmental issues are a natural extension of health and safety ones with regard to union activity, and points out that the Health and Safety at Work Act is one of the most powerful pieces of legislation regarding union involvement at the workplace.

The rights which safety legislation gives to union safety representatives are absent from all other areas of union activity. Because of this position in the UK, the NCU along with other unions therefore believes that "safety representatives should take on responsibility for environmental issues in the workplace".

It sees safety representatives well placed to deal with the issues because of experience of dealing with health and safety, where there is overlap with environmental problems to some extent. Also safety representatives have rights to access to information on hazardous substances, for example, particularly vital in view of the lack of access to information on environmental matters.

The union sees environmental issues as an area for co-operative rather than conflictual forms of industrial relations and says:

> "Active trade union involvement in environmental protection requires a new approach based on cooperation, partnership and joint working between union and employer. That approach, strongly supported by the TUC in no sense undermines the collective bargaining and other representative functions of the union; on the contrary it can only complement and enhance them."

The Manufacturing, Science and Finance union, MSF, which represents around 650,000 workers, including workers employed in the chemicals industry produced a policy statement which recognises that in the short term, the introduction of more stringent environmental protection policies could have an adverse effect on its members' job security.

However, it believes that companies will need to alter their operations to comply with stricter environmental legislation and consumer demand, and invest in new product development, identify new markets, "clean up" processes and re-train employees in order to survive.
MSF is also promoting the development of 'clean technologies' as a solution to environmental problems in industry. The union held a conference on clean technologies in the Summer of 1990.

It says that MSF has a crucial role to play in the process as it represents thousands of scientists, designers and technologists with particular expertise. The union campaigns for the establishment of a UK Environmental Protection Executive to oversee all aspects of the environment, directed at prevention of pollution, with an effective system of penalties for breaches of legislation.

NALGO, the union representing workers in local government, public services and recently privatised utilities has had a lengthy and varied involvement in raising environmental awareness and developing constructive initiatives. Doubtless this reflects the particular interests and competence of NALGO activists many of whom work in environmental protection, either in local government or in utilities with critical environment links such as the water, gas and electricity industries. Hence, NALGO have been campaigning for some time for the establishment of an Environmental Protection Agency (they provided the funds for the Labour MP Ann Taylor, who at the time was the opposition 'Shadow'

Minister for the Environment, to produce her book 'Choosing our Future' which sets out her plan for an Environmental Protection Agency.) The union set up an Environment Action Group in 1990 to co-ordinate NALGO's environmental campaigns and to develop policy proposals. NALGO's education department was amongst the first to develop and implement environment training courses. Finally, it is worth noting that NALGO has undertaken an environmental audit of its own head office in London. ("An Environmental Review of the National and Local Government Officers' Association." April 1992.) They are believed to be the first British union to have taken such an initiative.

Unions including NALGO, the Inland Revenue Staff Federation, IRSF, and the Civil and Public Services Association, CPSA have all been affiliated for some time to environmental pressure groups at national level.

UK unions are drawing on comparisons in Europe, and using examples of gains made by unions particularly in Denmark and Germany to press companies for the same agreements in the UK. The NCU point to the agreement between its sister union in Germany, the DPG and the Ministry of Posts and Telecoms. And MSF are attempting to negotiate an agreement with ICI which has been negotiated with the German chemical union, IG Chemie.

4.4 Attitudes of trade unionists to environmental issues

It is clear that national union and TUC policy on the environment has not been translated into Collective action at workplace level; and debate at union conferences shows that there is still some reluctance on the part of trade union members to "take on" environmental issues.

The issue of job security is obviously a major concern. The debate on affiliation to Greenpeace at the Civil and Public Services Association conference in 1991 reflected opening views of delegates. Whilst the resolution to affiliate was carried, several delegates opposed it on the grounds that Greenpeace campaigns for the scrapping of nuclear energy and therefore "for jobs to be taken away from CPSA members".

Whilst there are fears that in the long term environmentally damaging industries may be forced out of business, there is also fear that in the short term, improving environmental performance is costly and could lead to closures.

At a recent conference, one National Union of Mineworkers delegate said that at the colliery where he worked, there was considerable community opposition to sea pollution from colliery waste. But as union members, the workforce saw that other more costly methods of waste management would increase costs and could lead to closure, and hence felt unable to take any action on improving the situation.

The lack of access to information and lack of education and training on environmental issues are mitigating against union involvement. There has been a lack of guidance available to how to tackle issues; although practical advice and information is starting to be produced by unions.

While the results of exposure to harmful substances are often evident because of the effects on workers, the results of environmental pollution are not so immediately obvious. This may mean that workers do not view the pollution outside of the workplace as an issue appropriate to industrial relations, particularly where access to information is severely restricted.

There may be conflicts between health and safety and environmental protection which may have limited union involvement on environmental issues at workplace level. The obvious example of this is the control of exposure of the workforce using extraction ventilation resulting in exposure of the local population instead.

This conflict has also been identified in attempting to use alternatives to environmentally damaging products, such as chlorofluorocarbons which were used in the manufacture of some insulating materials. Alternatives given as examples by the environmental pressure group Friends of the Earth were more hazardous from a health and safety point of view for workers handling the materials.

Examples of industrial action by union members has highlighted the concern felt on environmental issues, particularly related to safety concerns. For example, successful campaigns have been mounted against the use of a number of pesticides, notably DDT; and the use of hazardous substances, including asbestos.

Seamen and transport workers have refused to handle nuclear waste destined for disposal in the North Sea; and trade unionists have refused to allow the importation of toxic waste destined for incineration in the UK.

However, it is unclear how trade unionists see trade unions as the vehicle to convey environmental concerns, and how far they see environmental issues as an issue for the bargaining agenda.

Whilst the LRD survey suggests that, at least in the public sector, trade unionists are raising environmental issues at bargaining forums, particularly safety committees, the extent to which this is happening in general remains unclear.

A survey by Touche Ross suggests that employees feel less empowered to influence their firms environmental policy, though Touche Ross argue that this is changing and that under a future Labour Government environmental matters could become a much more important aspect of industrial relations.

4.5 Interaction with environmental activists and community organisations

In the LRD survey, only 4% of respondents indicated that green groups had been involved in implementing environmental improvements at the workplace. Greenpeace and Friends of the Earth were the groups which had been involved. However, guidance from green groups was being used in the drawing up of environmental policies, particularly by local authority employees.

The relationship between green groups and industry tends to be adversarial, with company's poor environmental record being publicised, particularly by Greenpeace. However, the launch of "Greenpeace Business" claims that good practice as well as bad will be highlighted.

Traditionally, the relationship between unions and green groups has also tended to be adversarial, particularly because of the "jobs versus environment" conflict. However, unions and green groups are attempting to work together, and alliances have been formed.

At national level, a number of unions have affiliated, and therefore give financial support to Friends of the Earth: Fringe conferences have been organised at the TUC where union leaders and environmental pressure group members have shared a platform. It is perhaps worth mentioning that health and safety officers from the largest two unions, GMB and TGWU have taken up posts in environmental pressure groups, Friends of the Earth, and World Wide Fund for Nature.

At local level, a group financed by Greenpeace, Communities against Toxics (CATS) has been set up to co-ordinate action across the country where local communities are campaigning against industrial pollution. Local trade unionists are amongst the membership of CATs.

A 1989 report on trade unions and green groups by Greenpeace said that where political agreement does occur, it usually involves conflict rather than co-operation; although there are exceptions. The report highlighted areas where trade unions and green groups have been able to co-operate. The National Union of Seamen action against the sea dumping of radioactive waste had led to the development of personal contacts at the most senior level in the NUS and Greenpeace. In this case, Greenpeace said that the absence of a significant job threat, and the growing awareness of a possible occupational risk combined to produce successful co-operation.

Green groups and unions also worked in co-operation in campaigns opposing the privatisation of water and electricity services. The campaigns on toxic waste, pesticides and asbestos also lead to trade union and green group co-operation.

However the report also outlines problem areas experienced where co-operation was not found to be possible. These included particular unions' policies on nuclear power because of job preservation; TUC opposition to the Titanium Dioxide Directive (1983) because of perceived job and competitiveness threat to UK plants polluting the North Sea; and the close relationship between the unions and employers in opposition to radical action on power station pollution control to prevent acid rain.

The report outlines difficulties mitigating against co-operation between the two groups, including the job preservation issue and mutual mistrust, but concludes that co-operation is mutually beneficial, and is possible in areas including co-funding of research; jointly campaigning on issues and sharing information.

5. CURRENT ENVIRONMENTAL CONFLICTS AND THE INDUSTRIAL ACTORS.

5.1 Toxic waste disposal

The issue of toxic waste disposal, particularly the disposal of waste by incineration has become an area of increasing conflict. The importation of toxic waste from Italy and Canada destined for incineration in the UK caused public outrage in 1988 and 1989, and attracted considerable opposition from unions, green groups and local communities. The TUC has called for a ban on the importation of toxic waste into the UK.

However, waste disposal, including toxic waste incineration, is increasingly being labelled as a "green" activity. Companies including Shanks and McEwan, which has waste incineration operations, and Leigh Interests, have developed corporate environmental policies, and are promoted as green investments.

Investment fund managers, and financial analyst argue that waste incineration is a green activity because it gets rid of dangerous toxic waste safely. They point out that one of Friends of the Earth's founder members is now a director of one waste incineration company. Several companies operating waste incineration processes have been identified by financial advisors as standing to benefit from increasing environmental legislation and the way the consumer spends money.

Shanks and McEwan and Leigh Interests have been identified as "companies demonstrating a positive commitment to preserving the environment" and have been successful in greening their image. "Cowboy contractors" operating on the cheap have been blamed for the industry's past problems.

However, there have been sustained campaigns against operators of toxic waste incinerators by communities against toxics, consisting of people living in the vicinity of incinerators and concerned about the operations.

Local communities fear that there are adverse health and environmental effects from the incineration of hazardous chemicals, particularly dioxins produced when polychlorinated biphenyls are burned. Local councils and food manufacturers have opposed plans for the siting of new incinerators.

Union involvement has included dock workers refusing to unload imported waste, and six workers at Leigh Interests went on strike over health and safety problems in the company's laboratories, and inadequate testing of toxic wastes.

One local MSF branch in an area with toxic waste incineration operations says "the practice of incinerating large amounts of toxic cocktails" should be ended, and that "the method of toxic waste disposal should be reached democratically, fully taking into account the views of the waste disposal authorities and the local population".

At least one incinerator has been closed down, that operated by a company called Rechem in Bonnybridge in Scotland. There was massive opposition to the plant from the local community and findings of research carried out into the health effects on local people including a high incidence of twinning in the area. Although the Rechem workforce was unionised, local unions supported the closure of the plant in this instance.

The waste disposal industry claims that new legislation in the form of new partly enacted Environmental Protection Act and forthcoming European directives will force higher standards on the industry. But Greenpeace criticises the lack of enforcement officers to ensure that legislation is complied with.

In summary, waste disposal companies and their financiers argue that the process is safe and environmentally friendly in that it disposes of dangerous toxic waste; whilst unions and green groups, local communities, local authorities and food manufacturers argue that the safety of the process has not be proved, and that research and development in clean production methods is necessary.

Disposal of toxic waste - Case Study

A "Report on the controversy surrounding Leigh Environmental Services" (Part of Leigh Interests) was prepared for Greenpeace in 1988. The following outline is taken from this report, and is an example of a conflictual relationship between a waste disposal company and the community. (It is worth noting that this Case Study was used by NALGO in their Environmental Training Course Handbook: "Greenprint for Action" published in 1990.)

Leigh's operations in Walsall, in the West Midlands area of England uses a process called "Sealosafe" which involves solidifying waste after it has been pumped into mineshafts. Residents in the area formed a group, Community Action against Toxic waste. The aim of the campaign was to drive Leigh out of Walsall.

Initial complaints began in the late 1960's as a result of foul smells emanating from mineshafts where chemicals and other waste products were being dumped. The situation continued for several years and worsened with incidents including one in May 1987, when the area was enveloped in a dense gas and school children were sent home suffering from stomach upsets.

1,800 local people signed a petition against the company and complained of headaches, migraines, nausea, vomiting, stomach upsets, diarrhoea, asthma, bronchitis, allergic complaints and miscarriage (1 in 5 pregnancies in the area ends in miscarriage).

The petition and symptoms was sent to the local authority for investigation of Leigh's operations. The matter was also referred to the Secretary of State for the Environment. The residents group also carried out a survey and found high incidence of cancer and leukaemia. A local conservative councillor, commented on Walsall's reputation for being the unhealthiest place in Britain.

She referred to the high incidents of cancer and leukaemia in clusters around the waste disposal site and also spoke of an ex-waste disposal site employee who, after twelve years of work had developed throat cancer, and had three tumours and his larynx removed. She called for cross-party support for the closure of Leigh on health grounds (September 1988).

The major complaint against Leigh is negligence, and there is concern about the long-term stability of the Sealosafe system. The environmental health department of the local authority responded by carrying out tests and commissioning research into the increase in asthma deaths into the Walsall area.

The company has been prosecuted for breach of site license. The "outstanding number of prosecutions" and bad publicity surrounding Leigh lead the Caird Group to unload their shares abruptly. Leigh have been prosecuted for over-filling quarries and have taken out an injunction to prevent Environmental Resources Limited from making site inspections.

According to the report, Leigh maintain the attitude that there is no problem. The company hired a new public relations officer to improve the companies image.
"Residents find it ironic that the dot on the i of Leigh (on their logo) is a daisy, and that their lorries are painted environmentally-friendly greens and yellows".

There appears to have been no attempt during the years to meet with the protesters, picketers and residents to discuss their grievances. However, Leigh have sent out monthly neighbourhood newsletters to residents calling themselves "good neighbours", to which residents reacted with outrage.

The current situation is that Leigh continue to operate in Walsall as it is not proved that Leigh's activities are prejudicial to health. Unfortunately, the report does not outline the role of the unions in this instance; although the position of the local MSF branch in Walsall has been previously outlined.

5.2 Industrial polluters

Increased environmental awareness has resulted in hostility from local communities to the siting of industrial premises, particularly those thought likely to pollute. According to a survey in the Economist, companies that want to make potentially polluting investments invest large amounts of management time in building links with local people.

It points to the example of British Petroleum wanting to build an oil platform in Poole Harbour, a beauty spot in the South of England. The development director spent more than a third of his time trying to allay the worries of local people. The company carried out extensive environmental research, ran computer simulations to study the tidal flows and sand movements, and drew up six different options for public debate.

According to the Economist, "giving local people a feeling that they have some control over what happens to their environment is, astute companies realise, one way to win friends and planning permission".

The reaction to industrial polluters by communities depends on the relationship with the workforce. Workers do not necessarily live in the community around the workplace and the two groups may not share common interests.

An example of conflict in this area was the proposal to extend Stanstead Airport in the East of England. Local people opposed the development on the grounds of increased pollution; whilst local unions welcomed the developments of new jobs in the area. The amount of unemployment in an area is obviously an important factor.

Industrial pollution - Case study

The following is an outline of a cooperative situation in which a chemical company and the unions represented there took joint action to secure improvements and reduce the amount of pollution caused by its operations.

SCM Chemicals Ltd is situated on the south bank of the River Humber and discharged its effluent into the river. The company produces titanium dioxide pigments for the paint industry. The problem at SCM was the discharge by pipe of byproducts of titanium dioxide pigment into the river. This had taken place in the Humber by SCM and another company, Tioxide UK, by legal consent since the 1980's.

As a result the coastline had turned a reddy brown colour due to the high acid levels in the water. Local fishermen had noticed that shrimps and crabs had been affected by the pollution, and that fish were often deformed. Many of the workforce live in the area, so the unions were aware of the damage caused by waste to the local environment, but were in the difficult position of wanting to improve the environment while protecting jobs.

However, the concern was such that the workforce contacted the environmental pressure group, Greenpeace, and a meeting was arranged with senior management at SCM. Industrial relations at the company were generally good. The company admitted that pollution was a problem and agreed that the only real solution would be an acid-recycling plant, which would cost up to £20 million. The recycling system was developed on-site by SCM engineers and therefore the actual cost was less, £14 million.

The company paid for union officials to visit Brussels and Strasbourg to find out more about the subject at European level. Meetings were arranged with MEPs and environmental experts. New legislation in the form of EC directives which will be implemented in the UK by the end of 1992 added to pressures from the workforce for the company to take action.

SCM Chemicals is one of the companies that the GMB has approached with a view to signing the environmental agreement.

6. SUMMARY AND RECOMMENDATIONS

The shift in public opinion in the UK in favour of environmental issues has resulted in increased pressures on industry to improve environmental performance. This has augmented the increasingly stringent legislation on pollution control being initiated at European level, notably the recently introduced Environmental Protection Act 1990

But this general increase in environmental awareness has yet to be translated into collective action, and environmental issues do not appear to have yet pervaded the system of industrial relations to any considerable degree.

It is only relatively recently, within the last few years, that the unions have campaigned on the involvement of unions on all aspects of workplace environmental issues. Previously, campaigns tended to focus on specific issues, such as the banning of pesticides, or the disposal of nuclear waste at sea.

The second wave of union representation has been to campaign for collective green agreements which give unions rights of involvement in this area. However, workers and their trade union representatives have no legal rights of involvement in environmental issues at the workplace, and it appears that few, if any employers have signed collective green agreements and voluntarily allowed union involvement in all areas of the company's environmental impact.

Clearly, this situation is likely to remain whilst there are no legal requirements on employers to involve unions in environmental issues. There is political hostility to the involvement of trade unions in environmental issues on the part of employers, as outlined by the CBI, who see that this would lead to a broadening of collective bargaining agenda and lead to union involvement in areas such as investment and other areas of company decision making.

Whilst employers and managers are more aware of environmental issues than ever, recent surveys have shown that many employers have not addressed environmental issues in their company, have not carried out audits and have not developed environmental policies.

Again, although there are more pressures on employers to examine the environmental impact of the company, there are no legal requirements for employers to carry out audits, with or without the involvement of unions.

Trade unions do not have any rights of representation on bodies responsible for developing policy and formulating new legislation on environmental protection, as is the case on health and safety, where the TUC nominate two representatives to the Health and Safety Commission.

Information availability on a company's environmental record is limited as only a small amount of information has to be kept in public register, there is low prosecution activity, and the information is not computerised at present which further restricts access.
This means that whilst a company may claim to have a good environmental record, it is extremely difficult to find out, whether as an employee, or a member of the public if this is in fact true.

It has generally been green groups who have exposed companys' polluting activities. Currently there is no legal protection for employees who "blow the whistle" on poor environmental practice, which means that they could be dismissed for breach of confidentiality.

These areas need to be addressed if greater freedom to information on environmental issue is to be achieved.

There is a serious lack of training and education on environmental issues provided to employees which needs to be addressed by both employers - the CBI says that a corporate environmental policy will only be successful with full employee commitment and involvement - and by unions who are urging their members to negotiate collective green agreements giving rights of involvement in all aspects of a company's environmental impact.

The issue of re-training for employees who may lose jobs in polluting industries due to the closure of "dirty" workplaces has not been seriously addressed despite a large potential market for environmental industries, for example the manufacture of pollution control equipment being identified.

It is perhaps surprising that in view of the lack of access to information and lack of training and education that more co-operation between trade unions and green groups is not in evidence. However, it appears that the main obstacle to building alliances in this area is the "job preservation v environmental protection issue".

It is difficult to assess the attitude of trade unionists to environmental issues as it appears there have not been extensive surveys carried out in this area. However, it seems likely that the lack of legal rights on involvement in environmental issues at the workplace, and lack of training, education and information are reasons why national union policy has not been translated into collective action at local level.

In conclusion, the integration of environmental issues into the system of industrial relations is in the very early stages in the UK. It is unclear at present how far employers will enter into voluntary agreements with unions, and indeed how far unions at workplace level are attempting to negotiate on these issues.

7. References

An earthly chance- Labour's programme for a cleaner, greener Britain, a safer,
sustainable planet
The Labour Party 1990

Are ethical investors green?
Labour Research July 1991
Labour Research Department

Business and the environment series
Financial Times 1990-1991

Business and the 'green' employee
Employment Affairs Report
Number 41 August 1991
CBI

CBI Environmental Work
Environment means business - A CBI action plan for the 1990's

CBI
Corporate Environment Policies/Environmental Statements
2/5/91

Changing Corporate Values - A guide to social and environmental policy and practice
in Britain's top companies
R Adams, J Carruthers, S Hamilton
New Consumer April 1991

Cost of Chemical clean up bill hits £16m
Sharon Watson
GMB journal July/August 1990

Environmental issues and policy implications: Towards the white paper
Submission by the Trade Union Congress to the environment white paper April 1990

Environmental Data Services Reports 1991

The Environmental Protection Act 1990
HMSO

Green Britain or Industrial Wasteland
Edited E. Goldsmith and N. Hildyard
Polity Press 1986

Greening of industry - An international view
New Scientist June 16 1990

Greening the workplace - a TUC guide to environmental policies and issues at work
August 1991

Greening of industry - An international view
New Scientist October 89

Greenpeace Business newsletter June and August 1991

How green is my workplace?
Labour Research September 1989 Vol 78 no 9

Industry and the Environment
Financial Times Survey
April 21 1989

Industry and the Environment Cleaning Up - A survey
The Economist September 8 1990

Industry and the Environment
Financial Times Survey
March 16 1990

Industry Inches towards green goals
Julie Hill, Director of Policy for the Green Alliance
The Environment - A Special Report
The Independent 10 September 1991

Industrial Relations and the Environment: a new role for Trade Unions?
Employment Affairs Report
Number 39 April 1991
CBI

London Environmental Bulletin
London Scientific Services
Spring 1990 Volume 6/1

Report of the Controversy Surrounding Leigh Environmental Services
Maureen Plantagenet BA (Hons)
January 1988

The Environmental Audit
A green filter for company policies, plants, processes and products
John Elkington Sustainability UK 1990

The Green Audit S Heaton, Industrial Society Magazine December 1990

This Common Inheritance - Britain's Environment Strategy
HMSO 1990

Towards a Charter for the Environment
General Council Statement to the 1989 Trade Union Congress

Trade Union Congress General Council Report 1990
Section F: Environment

Trade unions and green groups - from conflict to co operation
A report for Greenpeace UK
May 1989

Trade unions and the environment survey
Bargaining Report 108 Labour Research Department July 1991

Trade unions and the environment
An interim report by Dennis Tangney for the Socialist Environment and Resources Association
May 1988

Trade Union policy documents
Action on the environment
MSF
General Secretary designate: Roger Lyons
64-66 Wandsworth Common North Side
London SW18 2SH
Tel: 081 871 2100

Control of Environmentally Damaging Substances (COEDS)
A proposal from the NCU
NCU and the environment (conference papers)
General Secretary: Tony Young
Greystoke House
180 Brunswick Road
London W5 1AW
Tel: 081 998 2981

"Green Works"
GMB
General Secretary: John Edmonds
22-24 Worple Road
London SW19 4DD
Tel: 081 944 7129

Employer Organisations
Chemicals Industries Association
Kings Building
Smith Square
London SW1 3JJ
Tel 071 834 3399

Confederation of British Industry
Centre Point
103 New Oxford Street
London WC1A 1DU
Tel 071 379 7400

Green groups
Association for Conservation of Energy
9 Sherlock Mews, London, W1M 3RH
Tel 071 935 1495

Campaign for Lead Free Air
3 Endsleigh Street, London WC1 ODD
Tel 071 278 9686
Pressure group campaigning against lead in petrol and elsewhere

Friends of the Earth
26/28 Underwood Street, London, N1 7JU
Tel 071 490 1555
Founded 1971, Staff 60, Membership 150,000, Income £1,685,000

The organisation is keen to work with trade unions and individual trade unionists and has done so on many issues including pesticides, asbestos and toxic wastes.

NALGO and IRSF are among those unions which have affiliated and donated money to Friends of the Earth, David Gee, its director was formerly the national health and safety officer with the general union GMB and is therefore in a good position to help
establish links with unions.

It has produced many publications, of which several, including The Environmental Charter for Local Government and Briefing Sheets on Environmental Auditing, would
be very useful for unions planning negotiations on environmental issues. It also publishes a quarterly magazine, Earth Matters.

There are around 280 local groups around the country actively campaigning on a variety of environmental issues, including air and water pollution, countryside protection and agriculture, urban regeneration, transport, resources and recycling, toxic wastes, tropical rainforests, overseas aid and trade and, the greenhouse effect.

Greenpeace
30/31 Islington Green, London N1 8BR
Tel 071 354 5100
Founded 1971, Staff 50, Membership 265,000, Income £4 million.
Greenpeace describes itself as an international environmental pressure group which
maintains complete independence from all political parties anywhere in the world.

It has a policy of non-affiliation with any groups, but individual trade unionists are encouraged to join and trade unions can subscribe and receive regular information on Greenpeace campaigns.

Greenpeace has worked with unions at local and national levels in the past, including the campaign against the sea dumping of radioactive waste and the transport of toxic waste. It provides speakers for union meetings, information and help with campaigns.

It has outlined future action which it believes can be taken by trade unions including:
* union members establishing links with green groups
* green group representatives being invited to attend conferences and provide information for trade unions;

* Safety representatives receiving training on environmental issues; and
* environmental policies and campaigns being given more attention and resources and support being given to green groups on national and international issues; and
* exchange of information with green groups on environmental concerns.

Socialist Environment and Resources Association
11 Goodwin Street, London N4 3HQ
Tel 071 263 7424
Founded in 1973 to campaign for "the socialist policies necessary to combat the fouling of the environment and the squandering of natural resources". It is smaller than FoE and Greenpeace but has done work on the "greening of unions" for many years.

It publishes New Ground magazine quarterly, to which trade unionists have contributed articles on environmental issues. It has many affiliated union branches and seeks more. It has produced surveys on trade unions and the environment and held conferences on the issue.

Transport 2000
Walkenden House 10 Melton Street, London, NW1 2EJ
Tel 071 388 8386

Transport 2000 was founded by unions and environmental groups to argue for more environmentally sensitive transport policies. It campaigns for better public transport, for pedestrians and cyclists and for transport in Europe. It welcomes union affiliation.

Womens Environmental Network
287 City Road, London EC1V 1LA
Tel 071 490 2511
NALGO's Women's Rights Committee agreed to affiliate to this organisation. Issues campaigned on include pollution from paper mills, dioxins and unnecessary packaging on goods.

Enforcement authorities Environmental Health Officers
They are located in the environmental health departments of local authorities. They are responsible for air pollution control of industries and workplaces not covered by the HMIP.

Her Majesty's Inspectorate of Pollution
Department of the Environment
Romney House
Marsham Street
London SW1 3PY
Responsible for "integrated pollution control" of prescribed industries under the Environmental Protection Act.

National Rivers Authority
Based in regional offices. It is responsible for the control of pollution in UK waterways.

INDUSTRIAL RELATIONS AND THE ENVIRONMENT:

CONTRIBUTORS

AUSTRIA.	by Dietmar Nemeth, Siegfried Jantscher Istitute fur Berufs-und, Erwachsenenbildungsforschung, an der Universitat Linz.
BELGIUM.	by Marc de Greef. ANPAT, Rue Gachard 88, BTE. 4 1050 Brussels
DENMARK.	by Borge Lorentzen, Kim Christiansen, Michael Sogaard Jorgensen. Interdisciplinary Centre Technical University of Denmark. Building 208 DK-2800 Lyngby, Denmark
FRANCE.	by Dr. Denis Duclos, Director of the Research Group: "Societies et Risques Technologoques" (SORISTEC) Centre National de la Recherche Scientifique. 16 Rue Moreau, 75012 Paris
GERMANY.	by Eberhard Schmidt. C. v. Ossietzky-Universitat, Oldenburg Inst. f. Politikwissenschaft II: Politik und Gesellschaft D-2900 Oldenburg, Ammerlander Heerstr. 114-118, 0441/798 2042/2639.
GREECE.	by Dr. Christina Theohari, Mr Ilias Banoutsos. Athens Labour Centre, Department of the Environment, 48B, 3rd September Str., 10433 Athens
ITALY.	by Dr. Alessandro Notargiovanni. C.R.D.A. (Centro Ricerche Documentazione Ambientale).Corso d'Italia, 25 00198 Rome.
NETHERLANDS.	by Drs C. G. Le Blansch. Centre for Policy and Management Studies Concept.Muntsraat 2A Rijksuniversiteit Utrech, 3512 E.V. Utrecht.
SPAIN	by Ernest Garcia (Coordinator), Rafael Gadea, Ignacio Lerma, Maria Luisa Lopez, Alicia Marcos, Jose Maria Ramirez, Antonio Santos Ortega. Department of Sociology and Social Anthropology University of Valencia.
UNITED KINGDOM	by Andrea Oates. Labour Research Department, 78 Blackfriars Road, London. SE1 8HF

European Foundation for the Improvement of Living and Working Conditions

Industrial Relations and the Environment
Ten Countries Under the Microscope
Volume II
(Reports on Greece, Italy, Netherlands, Spain and U.K.)

Luxembourg: Office for Official Publications of the European Communities

1993 — 250 pp. — 16 × 23.5 cm

ISBN 92-826-6023-0 (Vol. II)

ISBN 92-826-6021-4 (Vol. I and II)

Price (excluding VAT) in Luxembourg:

Vol. II: ECU 25
Vol. I and II: ECU 50

Venta y suscripciones • Salg og abonnement • Verkauf und Abonnement • Πωλήσεις και συνδρομές
Sales and subscriptions • Vente et abonnements • Vendita e abbonamenti
Verkoop en abonnementen • Venda e assinaturas

BELGIQUE / BELGIË

Moniteur belge /
Belgisch Staatsblad

Rue de Louvain 42 / Leuvenseweg 42
B-1000 Bruxelles / B-1000 Brussel
Tél. (02) 512 00 26
Fax (02) 511 01 84

Autres distributeurs /
Overige verkooppunten

Librairie européenne /
Europese boekhandel

Rue de la Loi 244 / Wetstraat 244
B-1040 Bruxelles / B-1040 Brussel
Tél. (02) 231 04 35
Fax (02) 735 08 60

Jean de Lannoy

Avenue du Roi 202 / Koningslaan 202
B-1060 Bruxelles / B-1060 Brussel
Tél. (02) 538 51 69
Télex 63220 UNBOOK B
Fax (02) 538 08 41

Document delivery:

Credoc

Rue de la Montagne 34 / Bergstraat 34
Bte 11 / Bus 11
B-1000 Bruxelles / B-1000 Brussel
Tél. (02) 511 69 41
Fax (02) 513 31 95

DANMARK

J. H. Schultz Information A/S

Herstedvang 10-12
DK-2620 Albertslund
Tlf. 43 63 23 00
Fax (Sales) 43 63 19 69
Fax (Management) 43 63 19 49

DEUTSCHLAND

Bundesanzeiger Verlag

Breite Straße 78-80
Postfach 10 80 06
D-W-5000 Köln 1
Tel. (02 21) 20 29-0
Telex ANZEIGER BONN 8 882 595
Fax 2 02 92 78

GREECE/ΕΛΛΑΔΑ

G.C. Eleftheroudakis SA

International Bookstore
Nikis Street 4
GR-10563 Athens
Tel. (01) 322 63 23
Telex 219410 ELEF
Fax 323 98 21

ESPAÑA

Boletín Oficial del Estado

Trafalgar, 29
E-28071 Madrid
Tel. (91) 538 22 95
Fax (91) 538 23 49

Mundi-Prensa Libros, SA

Castelló, 37
E-28001 Madrid
Tel. (91) 431 33 99 (Libros)
       431 32 22 (Suscripciones)
       435 36 37 (Dirección)
Télex 49370-MPLI-E
Fax (91) 575 39 98

Sucursal:

Librería Internacional AEDOS

Consejo de Ciento, 391
E-08009 Barcelona
Tel. (93) 488 34 92
Fax (93) 487 76 59

Llibreria de la Generalitat
de Catalunya

Rambla dels Estudis, 118 (Palau Moja)
E-08002 Barcelona
Tel. (93) 302 68 35
     302 64 62
Fax (93) 302 12 99

FRANCE

Journal officiel
Service des publications
des Communautés européennes

26, rue Desaix
F-75727 Paris Cedex 15
Tél. (1) 40 58 75 00
Fax (1) 40 58 77 00

IRELAND

Government Supplies Agency

4-5 Harcourt Road
Dublin 2
Tel. (1) 61 31 11
Fax (1) 78 06 45

ITALIA

Licosa SpA

Via Duca di Calabria, 1/1
Casella postale 552
I-50125 Firenze
Tel. (055) 64 54 15
Fax 64 12 57
Telex 570466 LICOSA I

GRAND-DUCHÉ DE LUXEMBOURG

Messageries du livre

5, rue Raiffeisen
L-2411 Luxembourg
Tél. 40 10 20
Fax 40 10 24 01

NEDERLAND

SDU Overheidsinformatie

Externe Fondsen
Postbus 20014
2500 EA's-Gravenhage
Tel. (070) 37 89 911
Fax (070) 34 75 778

PORTUGAL

Imprensa Nacional

Casa da Moeda, EP
Rua D. Francisco Manuel de Melo, 5
P-1092 Lisboa Codex
Tel. (01) 69 34 14

Distribuidora de Livros
Bertrand, Ld.ª

Grupo Bertrand, SA

Rua das Terras dos Vales, 4-A
Apartado 37
P-2700 Amadora Codex
Tel. (01) 49 59 050
Telex 15798 BERDIS
Fax 49 60 255

UNITED KINGDOM

HMSO Books (Agency section)

HMSO Publications Centre
51 Nine Elms Lane
London SW8 5DR
Tel. (071) 873 9090
Fax 873 8463
Telex 29 71 138

ÖSTERREICH

Manz'sche Verlags-
und Universitätsbuchhandlung

Kohlmarkt 16
A-1014 Wien
Tel. (0222) 531 61-0
Telex 112 500 BOX A
Fax (0222) 531 61-39

SUOMI/FINLAND

Akateeminen Kirjakauppa

Keskuskatu 1
PO Box 128
SF-00101 Helsinki
Tel. (0) 121 41
Fax (0) 121 44 41

NORGE

Narvesen Info Center

Bertrand Narvesens vei 2
PO Box 6125 Etterstad
N-0602 Oslo 6
Tel. (22) 57 33 00
Telex 79668 NIC N
Fax (22) 68 19 01

SVERIGE

BTJ

Tryck Traktorwägen 13
S-222 60 Lund
Tel. (046) 18 00 00
Fax (046) 18 01 25
          30 79 47

SCHWEIZ / SUISSE / SVIZZERA

OSEC

Stampfenbachstraße 85
CH-8035 Zürich
Tel. (01) 365 54 49
Fax (01) 365 54 11

ČESKÁ REPUBLIKA

NIS ČR

Havelkova 22
130 00 Praha 3
Tel. (2) 235 84 46
Fax (2) 235 97 88

MAGYARORSZÁG

Euro-Info-Service

Club Sziget
Margitsziget
1138 Budapest
Tel./Fax 1 111 60 61
         1 111 62 16

POLSKA

Business Foundation

ul. Krucza 38/42
00-512 Warszawa
Tel. (22) 21 99 93, 628-28 82
International Fax & Phone
(0-39) 12-00-77

ROMÂNIA

Euromedia

65, Strada Dionisie Lupu
70184 Bucuresti
Tel./Fax 0 12 96 46

BĂLGARIJA

Europress Klassica BK Ltd

66, bd Vitosha
1463 Sofia
Tel./Fax 2 52 74 75

RUSSIA

Europe Press

20, Sadovaja-Spasskaja Street
107078 Moscow
Tel. 095 208 28 60
          975 30 09
Fax 095 200 22 04

CYPRUS

Cyprus Chamber of Commerce and
Industry

Chamber Building
38 Grivas Dhigenis Ave
3 Deligiorgis Street
PO Box 1455
Nicosia
Tel. (2) 449500/462312
Fax (2) 458630

TÜRKIYE

Pres Gazete Kitap Dergi
Pazarlama Dağitim Ticaret ve sanayi
AŞ

Narlibahçe Sokak N. 15
Istanbul-Cağaloğlu
Tel. (1) 520 92 96 - 528 55 66
Fax 520 64 57
Telex 23822 DSVO-TR

ISRAEL

ROY International

PO Box 13056
41 Mishmar Hayarden Street
Tel. Aviv 61130
Tel. 3 496 108
Fax 3 544 60 39

UNITED STATES OF AMERICA/
CANADA

UNIPUB

4611-F Assembly Drive
Lanham, MD 20706-4391
Tel. Toll Free (800) 274 4888
Fax (301) 459 0056

CANADA

Subscriptions only
Uniquement abonnements

Renouf Publishing Co. Ltd

1294 Algoma Road
Ottawa, Ontario K1B 3W8
Tel. (613) 741 43 33
Fax (613) 741 54 39
Telex 0534783

AUSTRALIA

Hunter Publications

58A Gipps Street
Collingwood
Victoria 3066
Tel. (3) 417 5361
Fax (3) 419 7154

JAPAN

Kinokuniya Company Ltd

17-7 Shinjuku 3- Chome
Shinjuku-ku
Tokyo 160-91
Tel. (03) 3439-0121

Journal Department

PO Box 55 Chitose
Tokyo 156
Tel. (03) 3439-0124

SOUTH-EAST ASIA

Legal Library Services Ltd

STK Agency
Robinson Road
PO Box 1817
Singapore 9036

AUTRE PAYS
OTHER COUNTRIES
ANDERE LÄNDER

Office des publications officielles
des Communautés européennes

2, rue Mercier
L-2985 Luxembourg
Tél. 499 28 -1
Télex PUBOF LU 1324 b
Fax 48 85 73/48 68 17

3/93